Williams Particles

Alan Williams

Published by Lulu.com

ISBN 978-1-291-89733-3

Bubble Force

The existence of bubble force appears to lead to large particles including atoms.

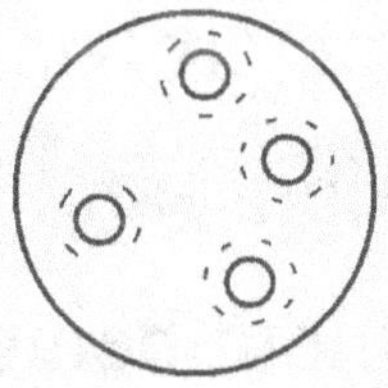

There are spheres. Portions of them expand.

Expanded portions cause pressure on a containing sphere.

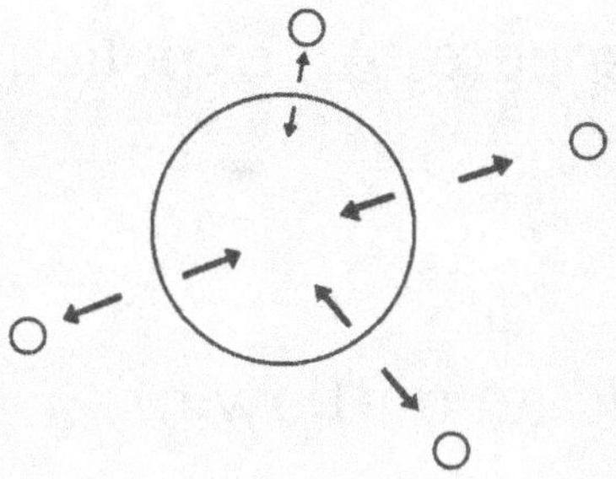

Some expanded portions break free.

In the opposite direction to each expulsion, a force is created.

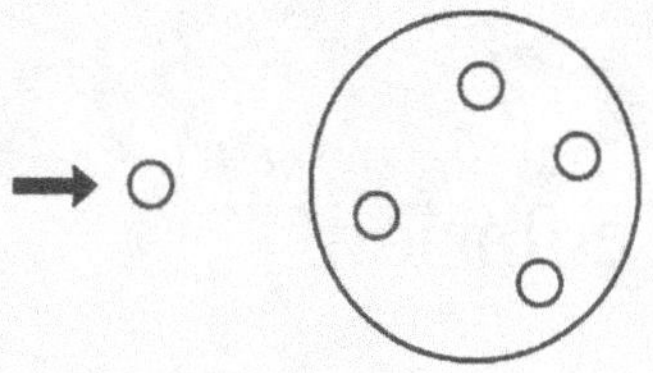

Expanded portions are able to strike a sphere from one main direction. A force travels through the sphere.

Some expanded portions are dislodged. Fired out to the right by internal pressure, they send the sphere to the left.

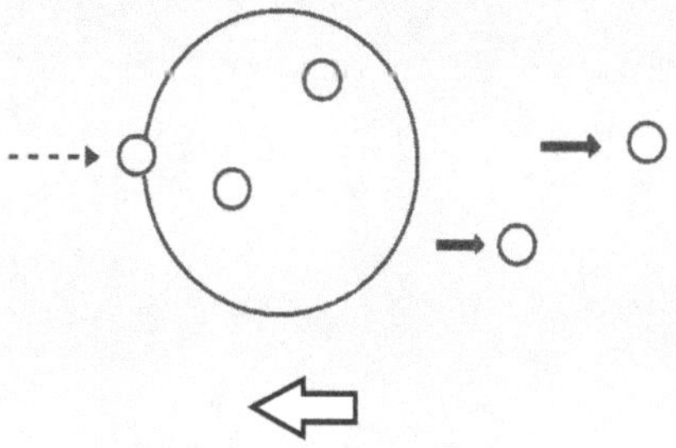

Saving a Sphere from Rapid Decay

A sphere consists of a huge number of very small particles, spherins. A spherin latches onto other spherins.

Small groups and individual spherins can enter a sphere. They are free to join with other unintegrated spherins. They create growing structures. These growing structures are expanded portions, and they create pressure against the containing sphere. Without being dislodged and sent out, expanded portions still tend to break free at high velocity.

An expanded portion is a small version of a sphere. An expelled expanded portion is inclined to grow in size.

At the edge of a sphere, there is an inflow and an outflow of material. This activity limits the growth of the structure.

Spheres sometimes break apart from a result of collision.

Forming a Sphere Column

Spheres attract other spheres by strikes with expanded portions. They form a travelling line.

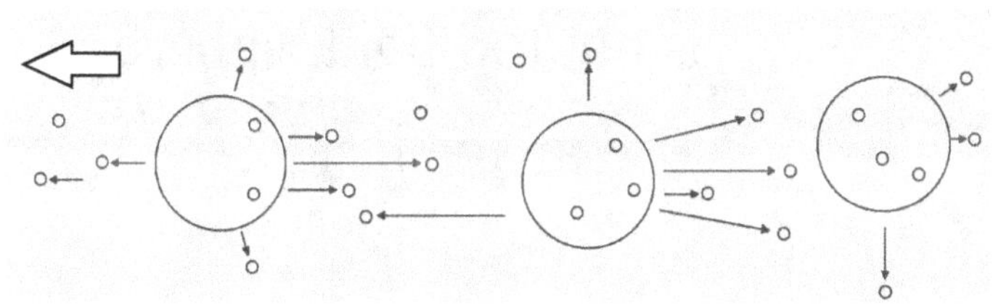

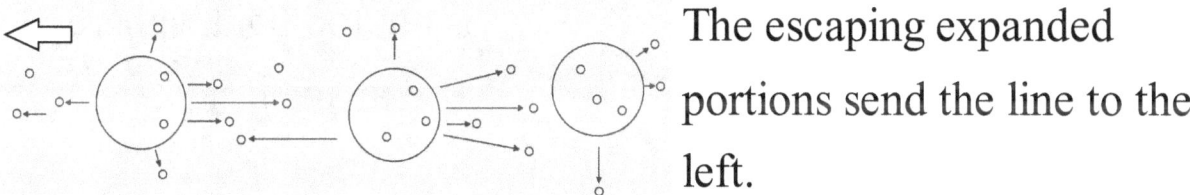

The escaping expanded portions send the line to the left.

The line attracts spheres alongside, and it can be imagined as a long travelling cloud.

The line is often knocked.

When the line of spheres has become sufficiently long, there is a chance for it to bend and for its ends to link.

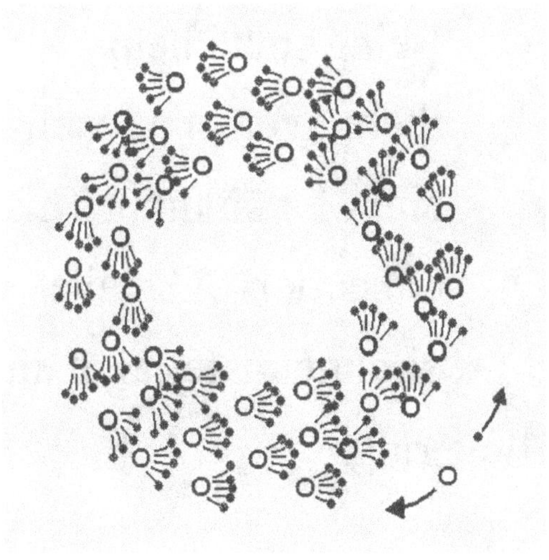

It becomes a circle of spheres. The expanded portions are represented by the black dots. In the diagram, expanded portions are causing spheres to circle clockwise.

Many expanded portions escape the circle, and they encourage other spheres to be nearer.

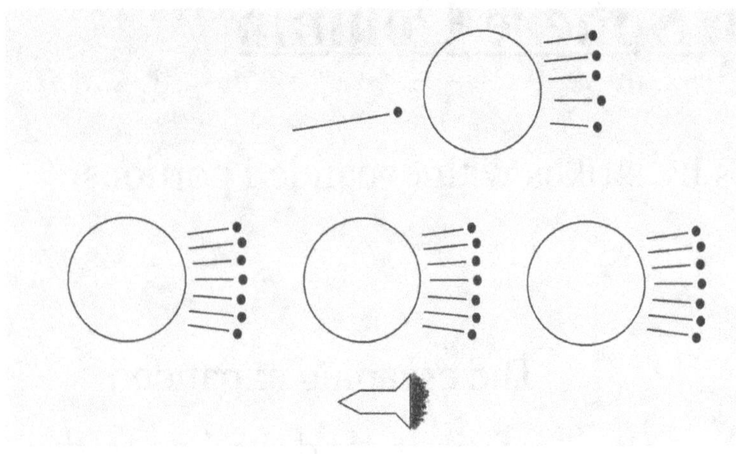

More spheres add to the circling. They may join above, below and surrounding the first circle of spheres. If the circling is not too tight, spheres also arrive inside.

When a circling sphere is struck at the front, some internal spherins are encouraged to the rear. This supply helps to maintain the circling.

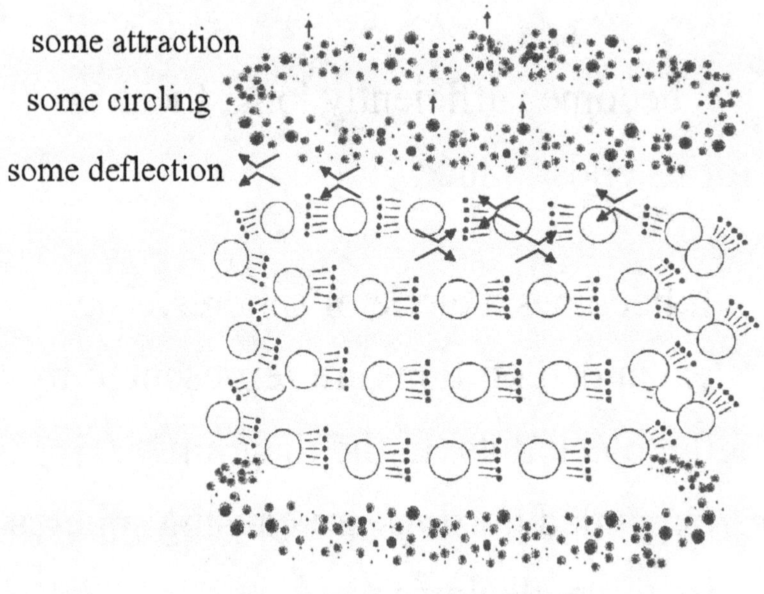

some attraction
some circling
some deflection

A column is formed. Expanded portions are often much smaller than spheres. Being small helps them to exist at each end of a column. This collection of smaller circling material at an end of a column can be called a small particle ring.

Circling expanded portions release spherins and small groups of spherins. In the circling action, expanded portions function in the same manner as spheres, and they circle in the same direction.

A growing expanded portion of a small particle ring may be struck by material from outside the structure. Before going far, the growing expanded portion is likely to be struck many times by material from the rest of the small particle ring. The structure of a small particle ring has some stability. Strikes from outside are able to influence an entire small particle ring.

Material of a small particle ring is largely occupied. With growth, the ring releases an increasing amount of very small particles, giving it an increasing field intensity. The small particle ring's distance from a column increases with its growth, until it is knocked away as a small particle ring.

When a column of spheres is complete, a sphere arriving very close to the column is frequently struck. The sphere rapidly loses internal expanded portions. Lacking these portions, the sphere is easily knocked away. It is not able to add itself to the column.

There is plenty of space for many expanded portions to escape through the centre of a column.

A sphere column is a long structure. Easily seen drawings of them do not always fit on a page.

For convenience, a sphere column can be represented by a brief structure.

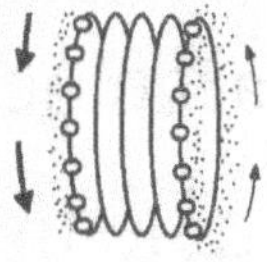

Attraction by Small Particle Rings

In phase (1), in an area of less bombardment, a travelling small particle ring increases its material growing internally. Expanded portions, having travelled through the centre of a column, strike the ring. At speed, the small particle ring heads towards the sphere column.

In phase (2), the arriving small particle ring begins to merge with the one forming at the column.

In phase (3), the arriving small particle ring strikes. Some expanded portions are sent downwards causing some upwards thrust.

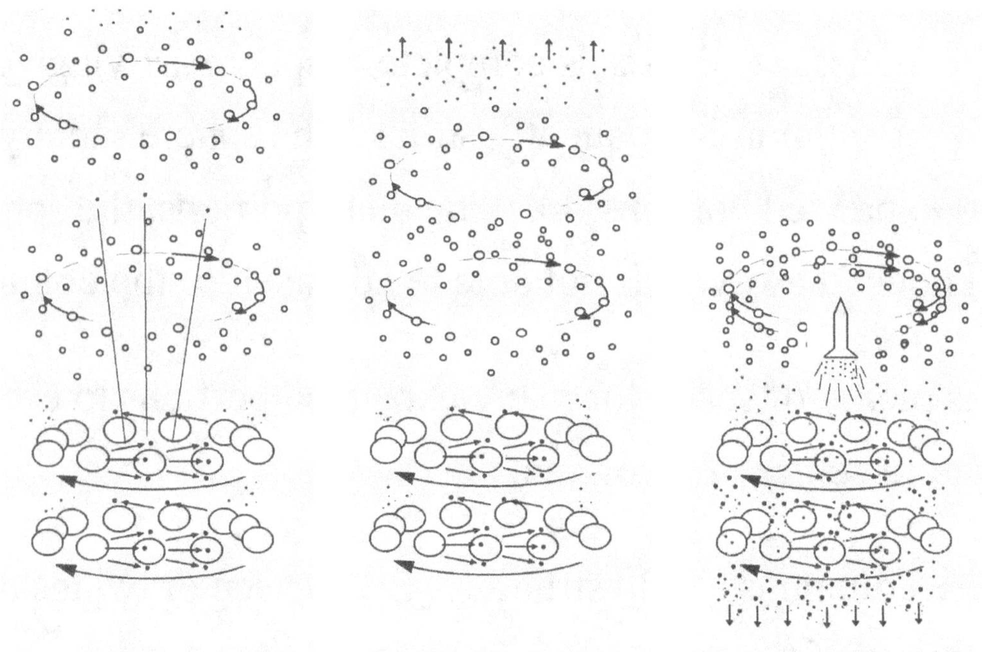

phase 1 phase 2 phase 3

Arriving small particle rings must have a sympathetic circling direction to cause an attractive force. They strike a side hard enough to release expanded portions at an opposing side.

A large occurrence of this process is able to move a much larger particle than a small particle ring.

Sphere Columns into Fundamental Core Structures

One sphere column encourages another to form at a distance.

Sphere columns send out small particle rings. Small particle rings are sometimes deflected.

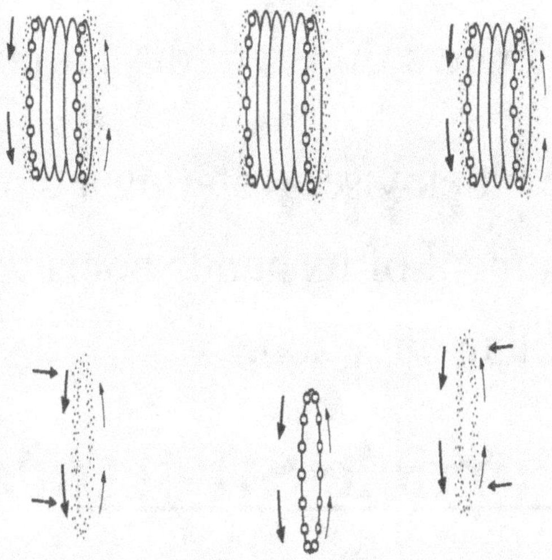

At a distance to a sphere column, more spheres are attracted to a side area.

Spheres naturally fall into circling. When the circling matches commonly arriving small particle rings, the circling spheres may remain and grow in number.

In the area adjacent to the sphere column chain, a sphere column begins to form.

In time, many sphere columns are next to each other. Each group is sending out spherins, expanded portions and small particle rings.

Growth is stopped by the busyness of the field.

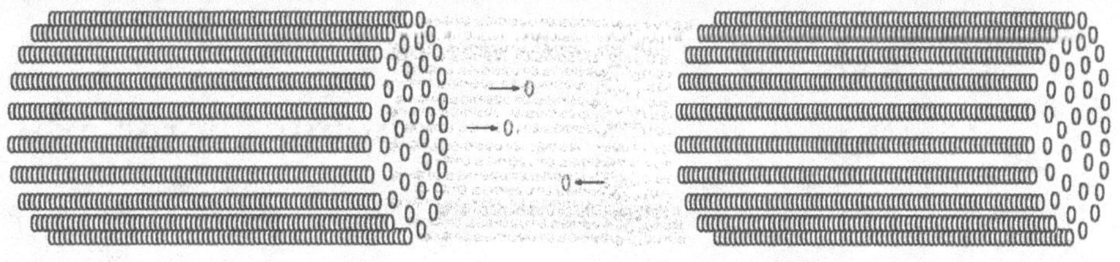

Added space makes this structure appear more natural.

Visible Circling Direction

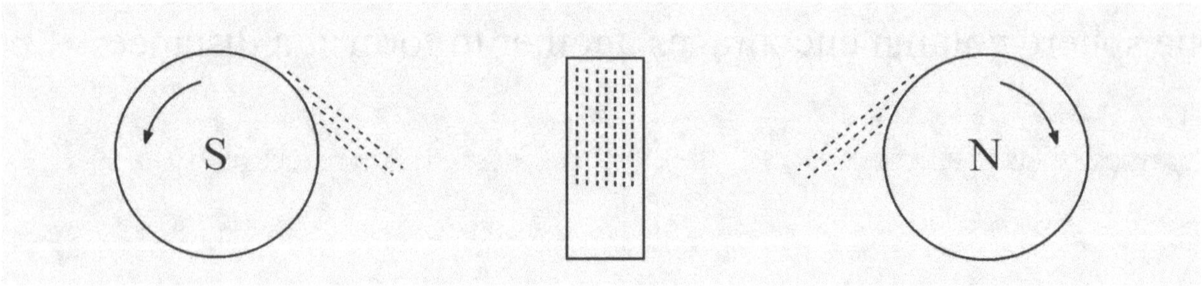
S
N

Although my eyesight is not able to see an individual particle, the term visible circling direction is usefully descriptive. The two ends of a spinning object spin in the same direction, but people moving up to the different ends see and experience a different direction.

N

North appears to be clockwise. Rotating a clock and noting the movement of its hands helps to illustrate visible circling direction.

Flux Line Particle Structures

Flux line particle structures are common, and they work well in a variety of circumstances.

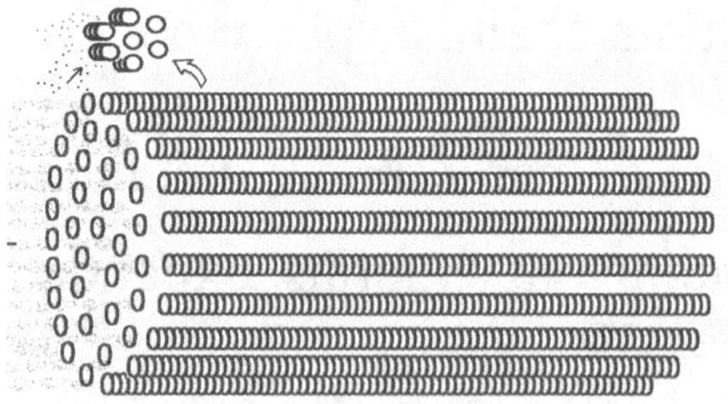

Expanded portions leave the side of a fundamental core, and they attract partially built cores.

These partially built cores are greatly restricted in size by the busyness of their environment. They grow into barrels.

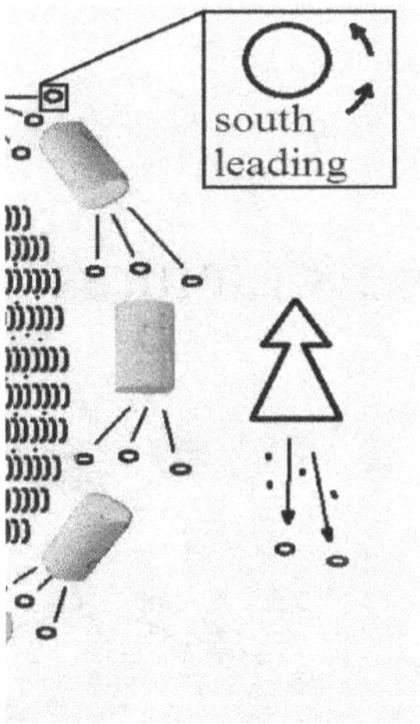
south
leading

Each barrel in the diagram is struck on its forwards end by small particles from the barrel ahead. Material is shifted to the rear. Small particle rings leave the rear. Expanded portions also leave the rear and propel the struck barrel along.

To fit with other areas, the small particle rings from barrels leave south visible circling direction, and they are south leading. (There might have been a contest at the very beginning.)

Below is a flux line particle structure. North leading small particle rings leave the north end.

N
NL
S
SL
SL
SL
SL

At south visible circling direction, the spheres at the core circle counterclockwise.

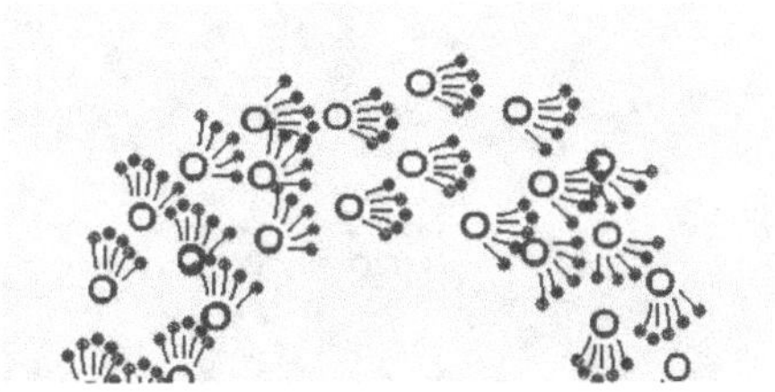

Still at south, the spheres send expanded portions in a roughly clockwise direction.

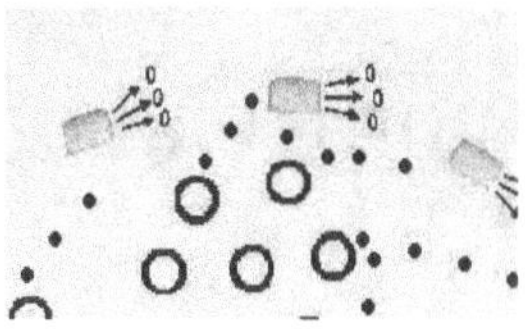

At an end, spheres in the core and barrels, they both circle in the same direction.

Forming an Electron

In a very high intensity environment, fundamental core structures do not gain disrupting barrels. Core structures attract other core structures. Many bond, and they form parallel lines. A heavily bonded group of core structures is an electron core.

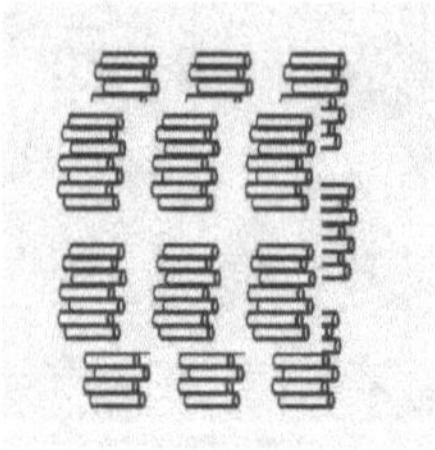

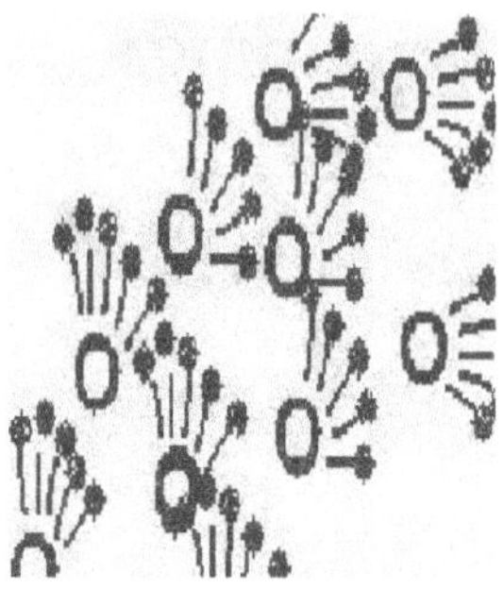

Expanded portions emanate from circling spheres of an electron core. Expanded portions attract flux line particle structures, and they give flux line particles their circling direction around the electron core. (A layer of barrels may exist.)

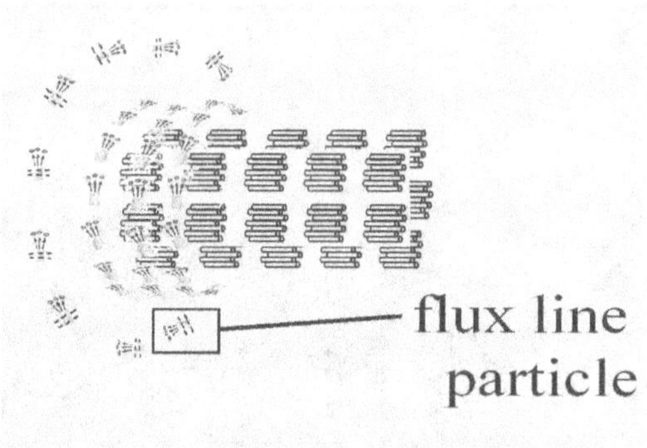
flux line
particle

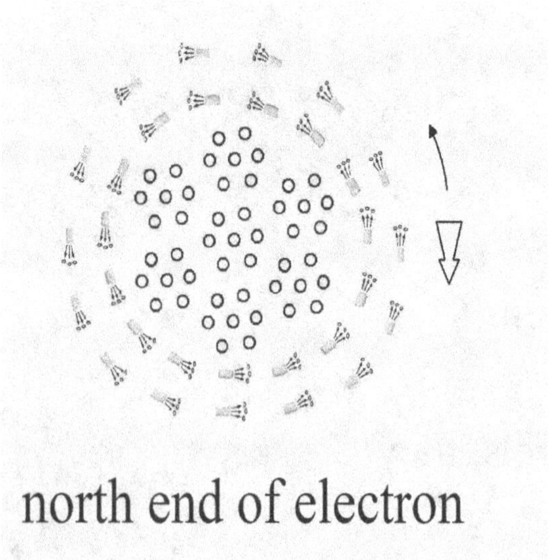
north end of electron

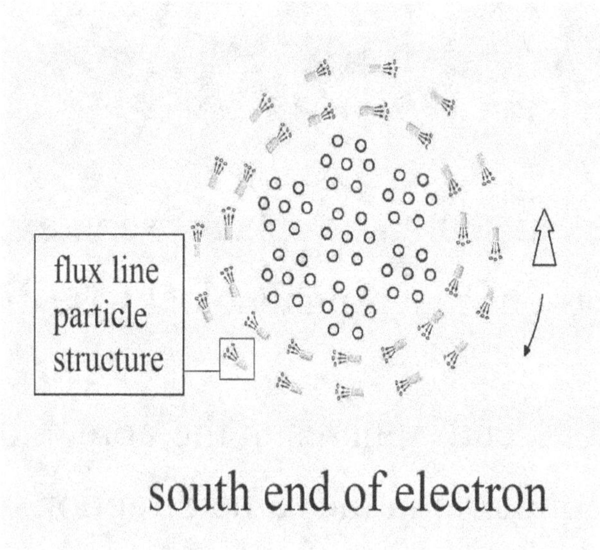
flux line
particle
structure
south end of electron

Between the flux line particles are many sphere column chains; these are not represented in the above diagrams. Sphere column chains connect the flux line particles. Circling flux line particles are increasingly compact towards the centre.

An electron can be held and orientated by flux lines. It is reasonable to believe an electron has a north end and a south end.

The flux line particles circling around electrons send out a large number of small particle rings. Having left the south end of flux line particles, the small particle rings are south leading. An early contest might have left these small particle rings being south leading. From material circling around a particle, south leading small particle rings are forming in a heavily bombarded area.

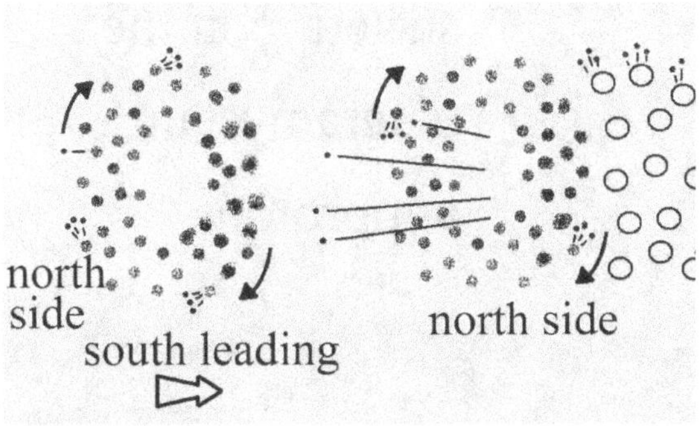

north side
south leading
north side

With a sympathetic circling direction, south leading small particle rings strike more cleanly and harder on a north side. The north side sends material to the south side.

The south end of an electron is less disturbed than its north end. The centre of the south end gains a protrusion. An electron can be represented by the following.

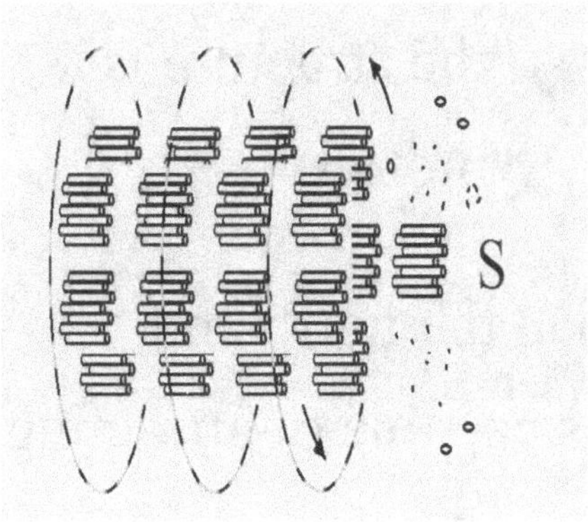

S

Material passing the protrusion is disrupted, and it is often deflected outwards in different directions.

Dashed lines represent circling flux line particle structures.

The relatively flat north end of an electron provides a much stronger field than its south end. Negative electric charge can be associated with the north field, clockwise visible circling direction.

Forming a Proton

If the environment is intense enough, the core of an electron does not gain flux line particle structures circling around it. With the aid of groups of concentrated flux lines, naked electron cores join and form parallel lines.

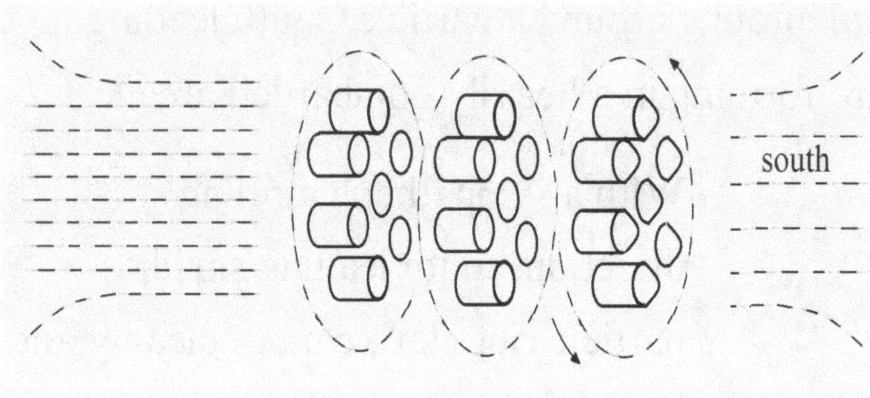
south

With its clear and flat north field, the structure in the diagram is an antiproton.

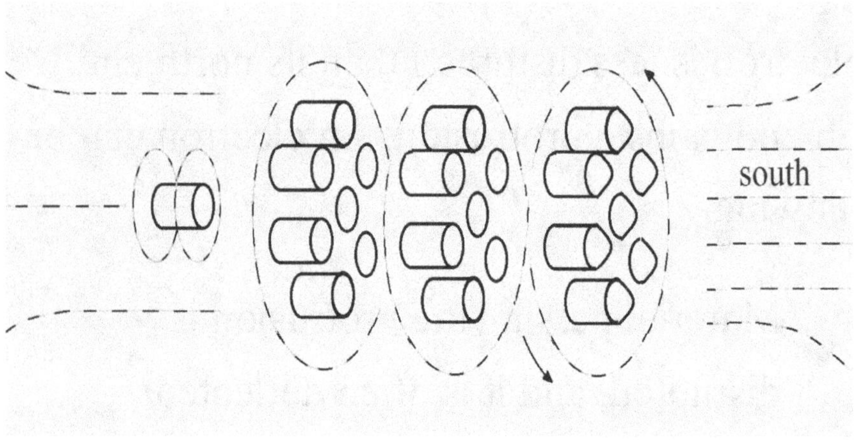
south

After the strong north field has secured material, the structure is a proton.

The field at the south end is now stronger than the north end. Depending on circumstance, the south end is capable of attracting an electron.

There are many circling flux line particles, and both ends of the proton have flux lines and groups of concentrated flux lines. Electron core structures at the south end of the proton are exposed, and each one has gained central material.

Positive electric charge can be associated with the south field, counterclockwise visible circling direction.

The Neutron

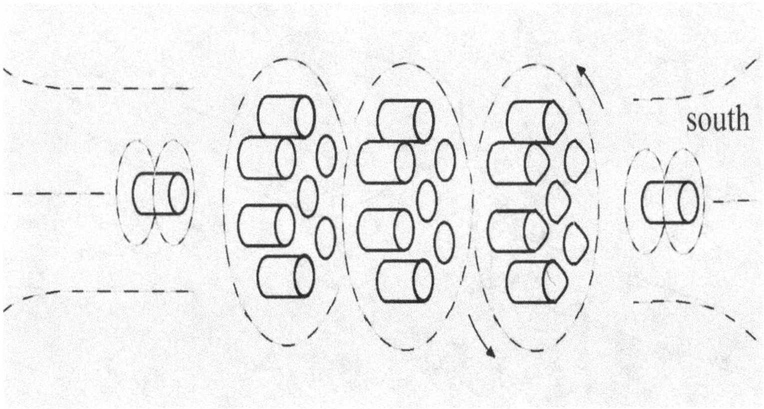
south

An isolated neutron is disrupted at both ends.

Unlike Poles Formation

North can be represented by the face of a clock, and south can be represented by the back.

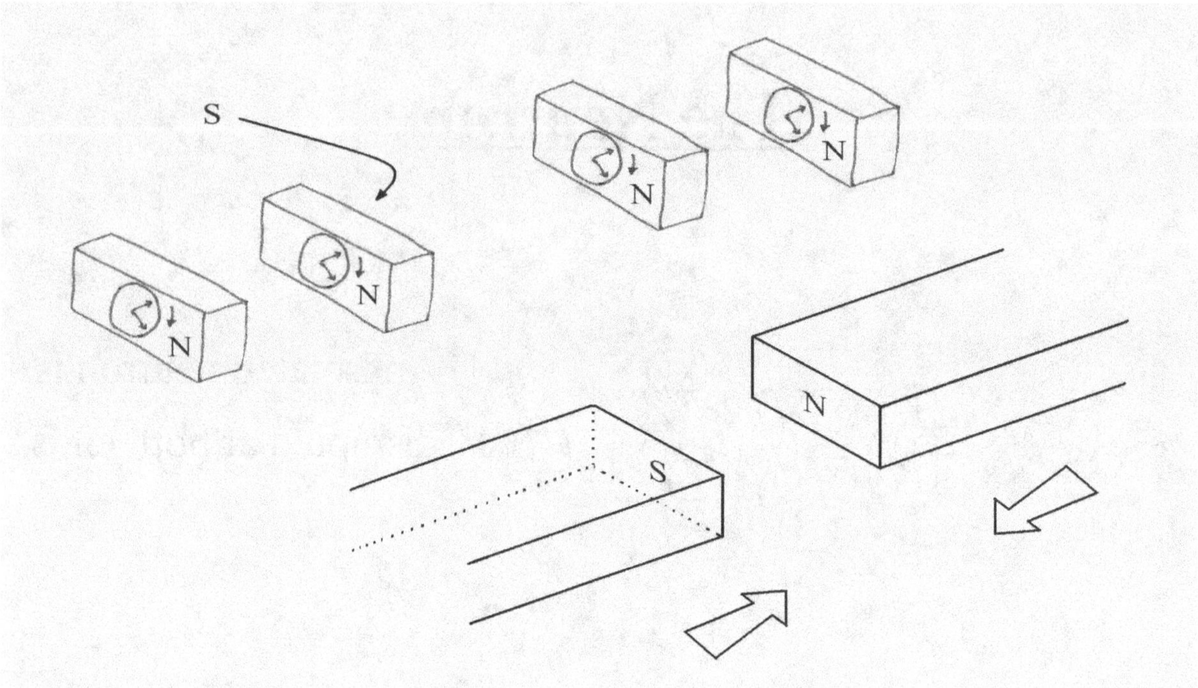

S
N
N
N
N
N
S

Depending on distance, unlike poles formation is attractive. Unlike visible circling directions, north with south, occurs with a common overall circling direction as represented by the clocks.

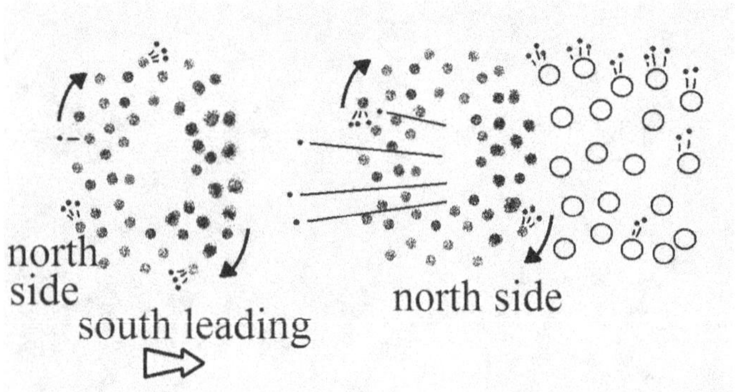

north side
south leading
north side

Arriving small particle rings merge sympathetically and strike. The result is an attractive force.

Repelling and Conflicting Like Poles

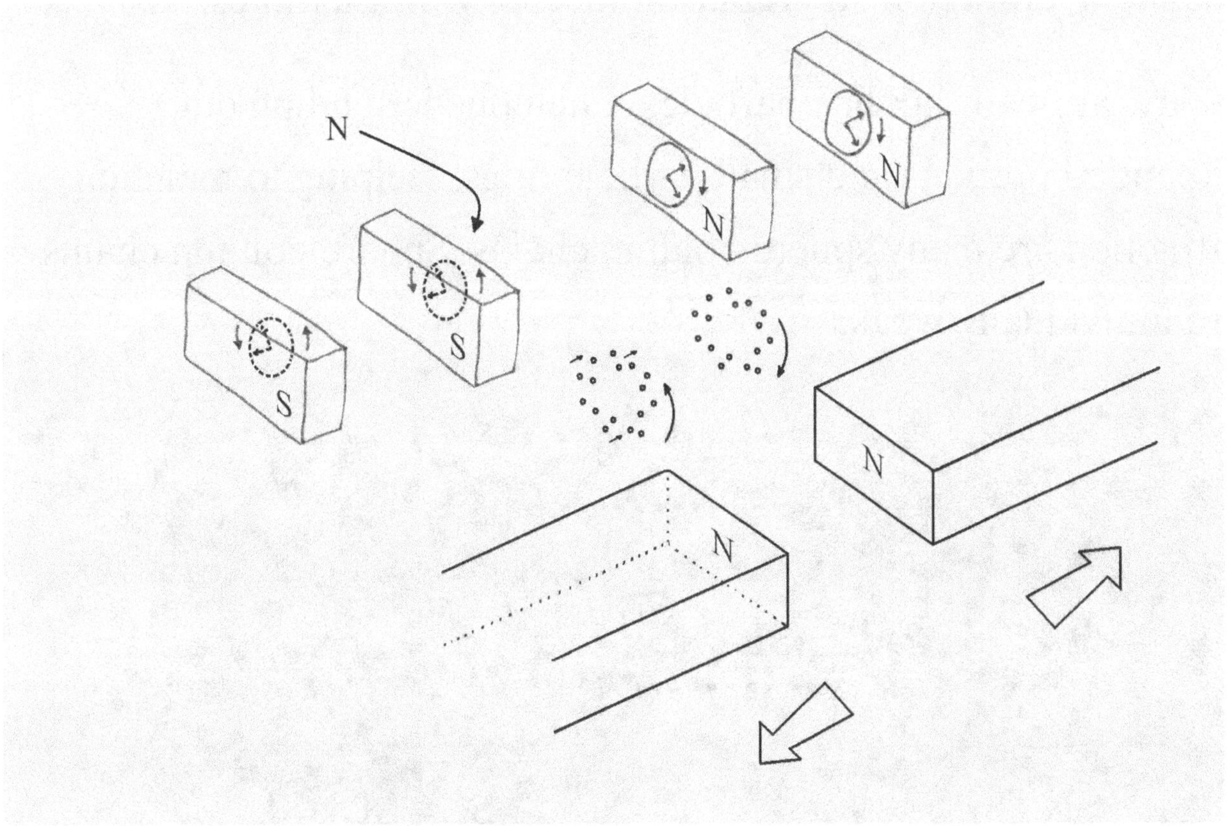
N
N
N
S
S
N
N

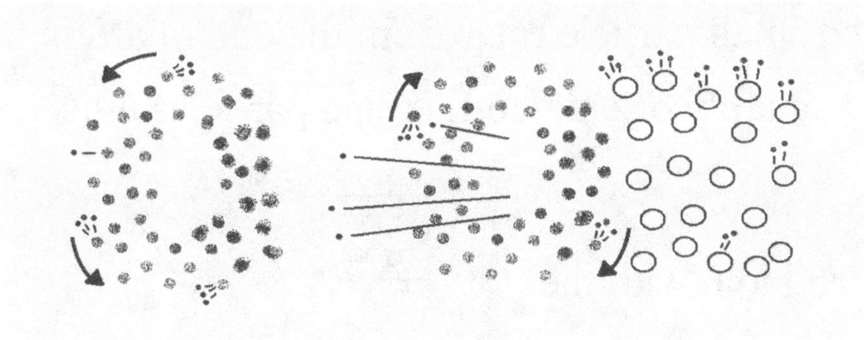

Many expanded portions escape through the centres of sphere columns. These expanded portions steer in small particle rings. When structures are in like poles formation, small particle rings from one structure to another do not harmoniously merge. Circling expanded portions head into each other, and some break apart. There are wild collisions. Expanded portions go in many directions, and there are many deflections. The result is a repelling force.

Magnetism

Magnetic attraction and repulsion occurs with flux lines.

Below are two flux line particles. Each one is sending out expanded portions and small particle rings. Helping to maintain attraction are many sphere column chains. Sphere column chains are individually weak.

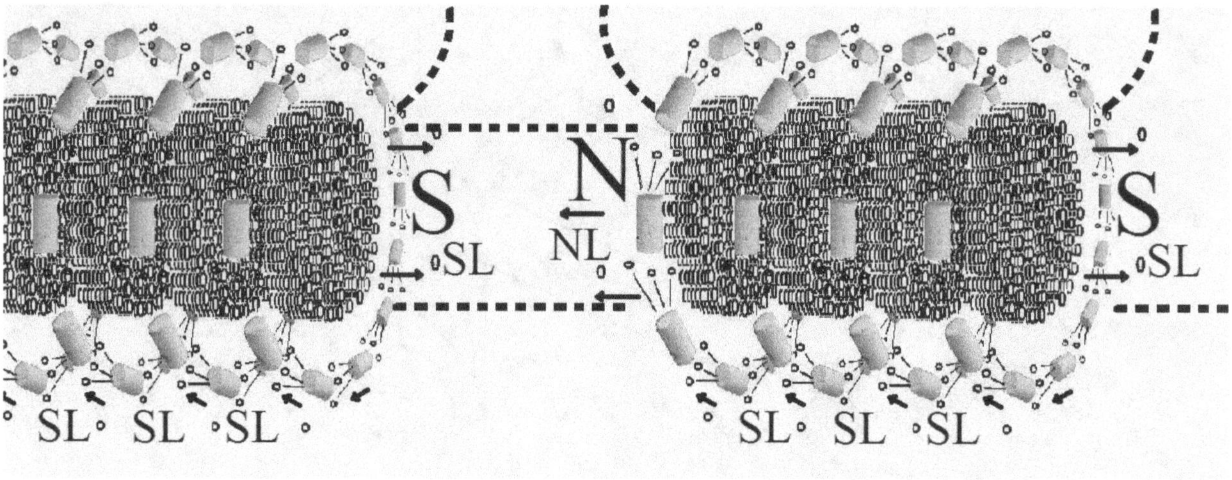
S
SL
N
NL
S
SL
SL SL SL
SL SL SL

Expanded portions and small particle rings from the end of a flux line are able to attract suitably orientated flux line particles. Flux lines grow quickly.

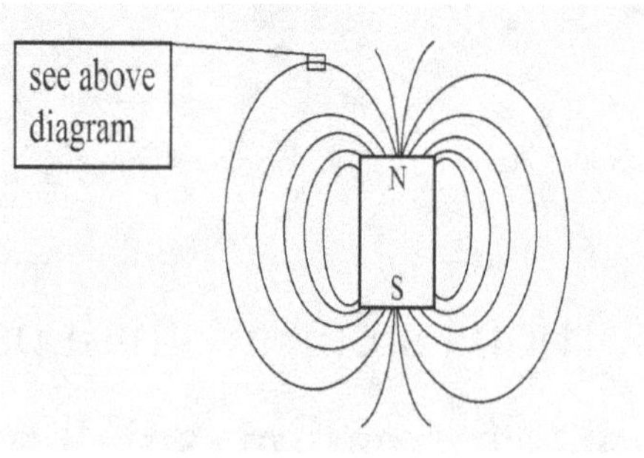
see above diagram
N
S

Many lines from one end join with the other.

Barrels, circling around flux line particles, provide some disruption, and they encourage alongside lines to be separate.

Near the face of a magnet and inside the material, flux line particles are bombarded more heavily. The bombardment reduces the numbers of barrels. This loss helps lines to be close together.

Flux lines are able to join with fundamental core structures of larger particles. For a flux line to join with a larger particle, the flux line must achieve close proximity.

Due to structure, only a few materials are able to join securely with the flux lines of a magnet.

From a magnet, flux lines are able to join with a small block of iron.

If there is a large enough number of flux line particles in a joined line, then the particles are approximately separated by field intensity. Field intensity is maintaining separation, when particles in a line are as much pushed away by outgoing smaller particles as they are drawn in by outgoing smaller particles. In this state of balance, particles in a line are relatively easy to knock away. While some particles can be knocked a little away and then drawn in again, the loss of some particles can be expected. With the loss, attraction is increased.

In the case where our small block of iron moves nearer to the magnet: flux lines reduce in length; particles in a line are closer together; and attraction and repulsion is more balanced. With this balance, there is another reduction in flux line particle numbers.

The iron, being closer to the magnet, helps flux lines to disrupt, invade and latch on in larger numbers; with these actions, the attraction between the iron and the magnet increases.

Flux lines from Earth are recognised to reach outside of our atmosphere. There is no apparent restriction to the growth of a free end. This achievement is understandable with the lines being of particles.

The Copper Tube Experiment

In the next diagram, a magnet and a piece of metal have the same shape, size and mass. Independently, they are dropped down the same two metres long copper tube.

Similar has been performed. The following has the poles of the magnet at the same height. The same result is expected.

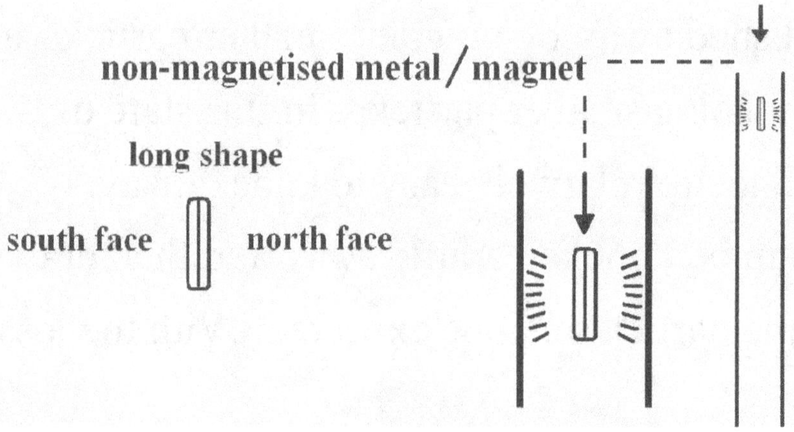
non-magnetised metal / magnet
long shape
south face
north face

The magnet takes the longest to fall. As it travels down, many weak bonds are formed and then broken. The time taken to fall can be used as an approximate measure of the often unnoticeable attraction existing between a magnet and copper metal.

Gravity

Gravity is a result of strikes from expanded portions and spherins. Also influential are small particle rings and temperature. We are able to consider the net effect of all these small particles, and we have the freedom to refer to them as gravity particles.

Gravity particles from the other side of Earth do not have to arrive directly. Their strikes often cause more particles to be released. The large numbers involved help a consistent level to reach us.

From a solitary ball of cheese, we are able to consider a moment's release of gravity particles.

The particles arrive at a metre away from the cheese. These gravity particles are alongside each other. We can ignore their length, and we can regard them as occupying the surface area of a sphere.

A sphere has a surface area of $4\pi r^2$.

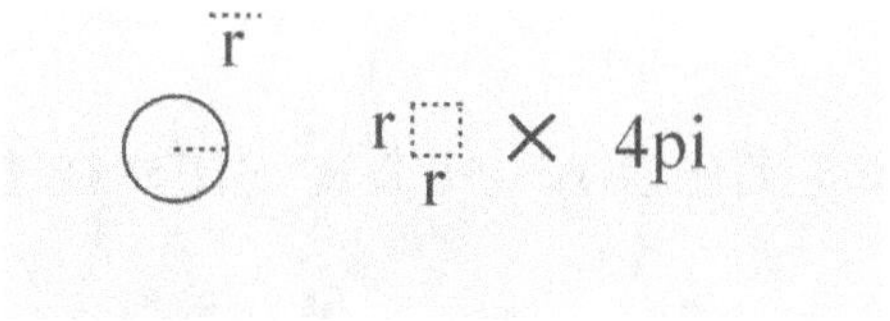
r
r
r
× 4pi

We are able to double the distance from the ball of cheese. 'r' becomes '2r'.

Substituting 'r' with '2r' gives

$$4\pi \times (2r)^2 = 4\pi \times 4r^2 = 4 \times 4\pi r^2.$$

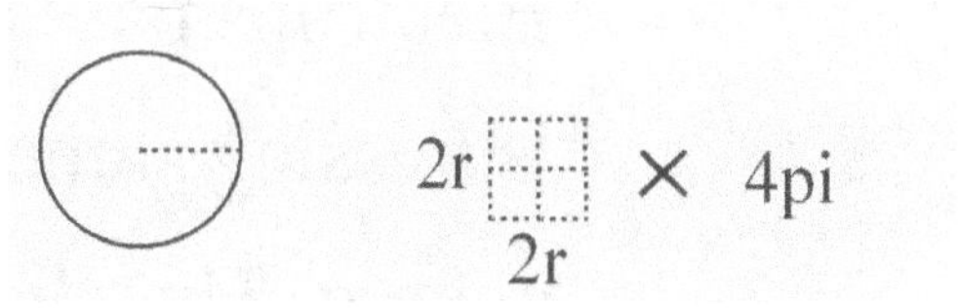
2r
2r
× 4pi

The second surface area is larger by a factor of 4. At two metres away from the cheese, the gravity particles must be four times more spread out than at the one metre distance. This spread is consistent with observation. Gravitational attraction quarters each time distance is doubled.

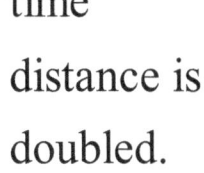

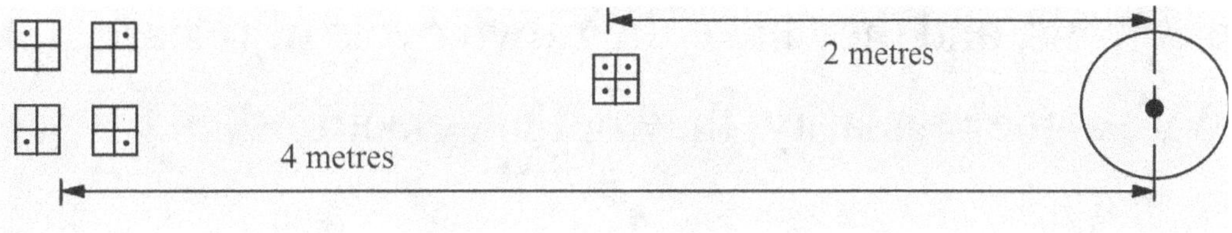
2 metres
4 metres

A Mystery Solved with Small Particles

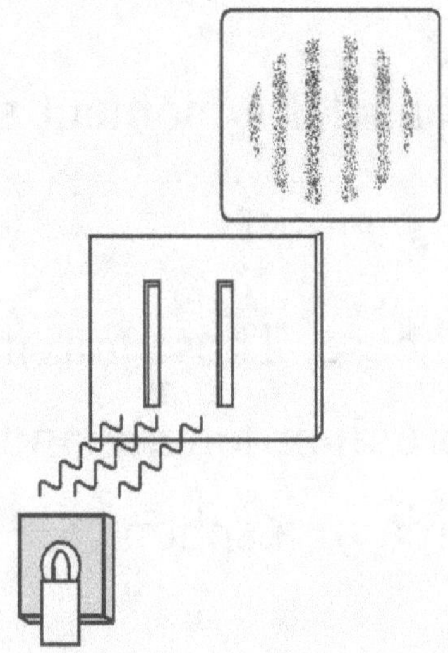

A single frequency of light reaches the two slits. Light waves encounter gravity from the slits, and many are deflected into a merging course. The waves interact, and they create an interference pattern. There is a case for a kind of coinciding wave deflection playing a larger role than destruction.

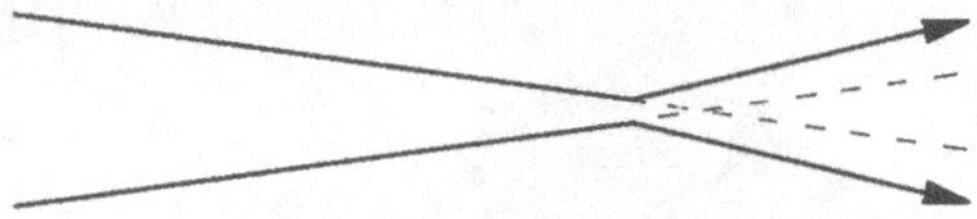

This idea will be mentioned again.

Blocking one of the slits brings a halt to the pattern.

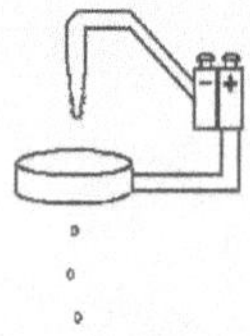

To the left, electrons are fired out one at a time. Recording results creates an interference pattern.

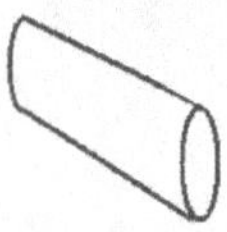

Electrons pass on either side of the acetate bar. The bar is missing some electrons; it has a positive electric charge. The charge encourages electrons to curve round the bar.

Flux line particle structures take both routes. Flux line particle structures are not always organised as in magnets. They depart from disturbed electrons, and they have their own cycle of release. With some regularity, they deflect electrons.

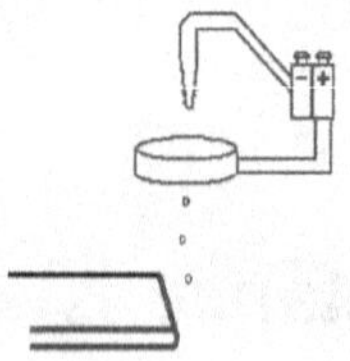

Blocking one route stops the pattern occurring.

The Electromagnetic Wave Propulsion System

In Order of Decreasing Wavelength
Radio Waves One wavelength can be over 1000 metres.
Television Waves
Radar Waves
Microwaves
Infra-red Rays
Visible Light Just one millimetre is about 1800 wavelengths.
Ultra-violet Rays
X-rays
Gamma Rays
Cosmic Rays There are well over 10 000 000 000 wavelengths to a millimetre.

Flux line particle structures take time to build and release small particle rings. At an end of a flux line particle structure, small particle rings can spend a little time being tenuously attached. Frequently, when one small particle ring is knocked away, others nearby are disturbed and released. A flux line particle structure on its own falls into having a periodic release.

Many flux line particle structures are able to form a wave.

Electromagnetic waves travel in a straight line. When waves go into a different and known clear material at an angle, they bend by a predictable amount. If a wave were to travel up and down, its angle of refraction would be more tasking to find.

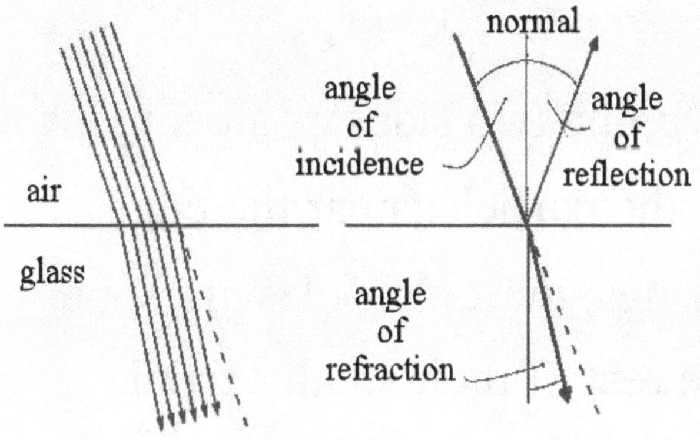
air
glass
normal
angle of incidence
angle of reflection
angle of refraction

Turning a corner requires slowing down on one side. From air to glass, the side of a wave arriving first reduces its speed before the other side.

At a time of low release of small particle rings, a wave particle's existence is much harder to notice.

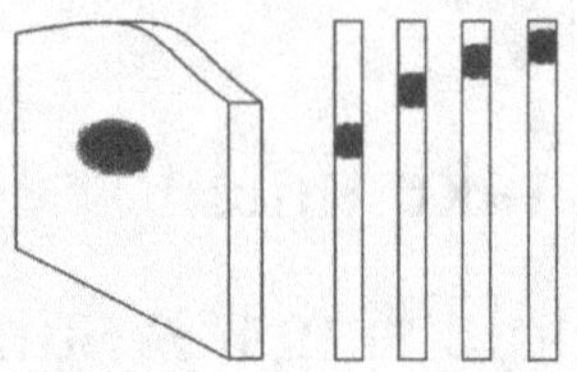

A wave has burnt a patch through some slices of bread. Due to their releasing phase, wave particles have an appearance of travelling up and down.

A recently discovered particle, the Higgs boson, made a brief vanishing act at the Hadron Collider. There was interesting speculation of it spending time in another dimension. More likely, it was having a rest from sending out particles.

A black rectangle can be used to highlight a particle in a releasing phase.

Imagine a group of wave particles travelling horizontally. The

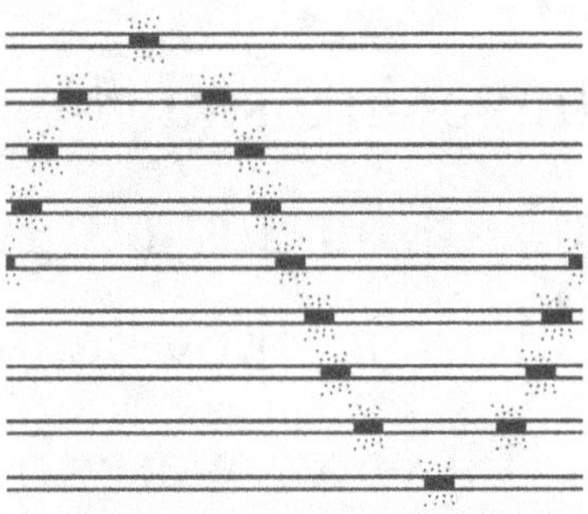

lowest wave particle goes into a releasing phase. This occurrence disturbs and triggers near and slightly higher wave particles into their releasing phase. This triggering action repeats. After the uppermost is releasing, the triggering action begins to work its way downwards. A wave shape is achieved.

The particles near the edge of a wave, the two sides of a peak, may be different particles. Alternatively, the particles near the edge release some of their small particle rings; these particles are soon disturbed again, and they release the rest of their small particle rings.

A wave also has width from particles being alongside.

In the diagram, flux line particle structures are travelling towards the right.

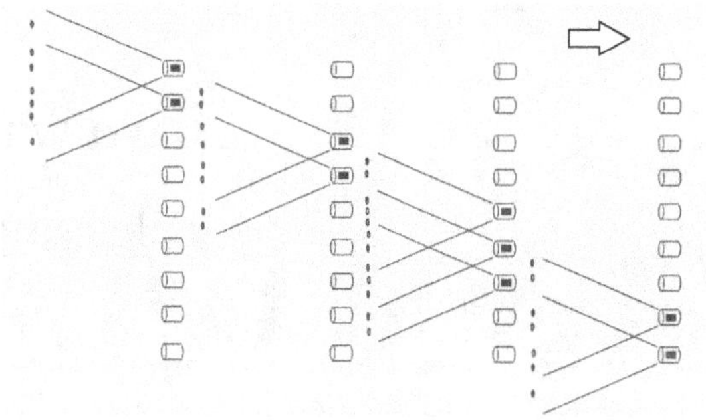

The forwards end of a wave particle is struck by material including small particle rings. Material, being sent to the rear, often dislodges more small particle rings and expanded portions. Propelling the wave particles along are the dislodged expanded portions.

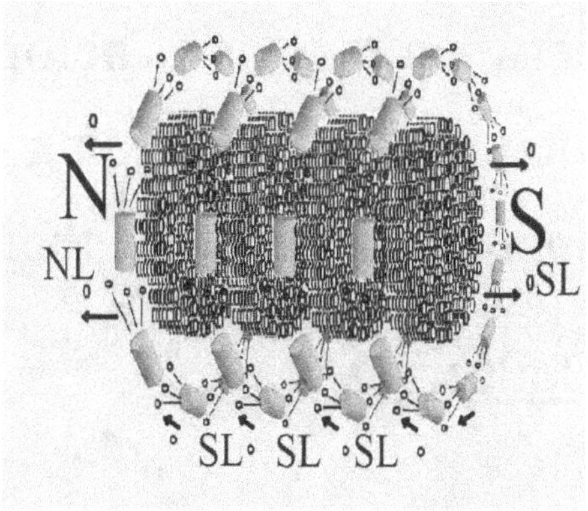
N
S
NL
SL
SL SL SL

Flux line particles moved into a busier environment lose some circling barrels. The loss of barrels helps compactness.

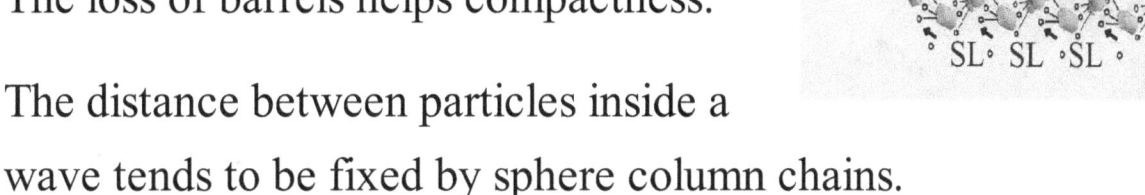

The distance between particles inside a wave tends to be fixed by sphere column chains.

Inside a wave, some particles may have an inclination to go one way, and others may have an inclination to go a different way. Being a very large number of connected particles, their direction averages out to straight ahead, unless there is a significant influence.

Constructive Waves

The visual result of the two-slit experiment does not prove the existence of constructive waves by itself. The following assumes they do exist.

Along an electromagnetic wave, each cycle has its own group of heavily releasing particles. Identical waves are able to interact in more than one area at the same time.

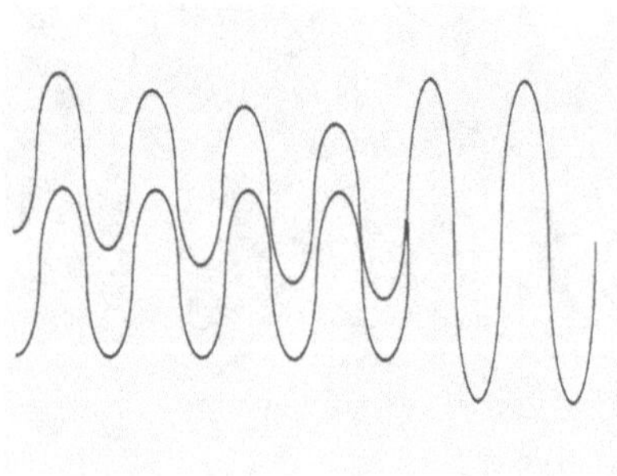

Two waves are constructive when they are releasing up and down in the same area.

Heavily releasing particles of one wave are amongst the heavily releasing particles of the other. Pressure amongst wave particles reduces by a spreading out in the direction of amplitude. Particles spread until the distance between them is as before the meeting. With a little time, two identical and coinciding waves combine and possess approximately twice their previous amplitude.

Deflecting Waves

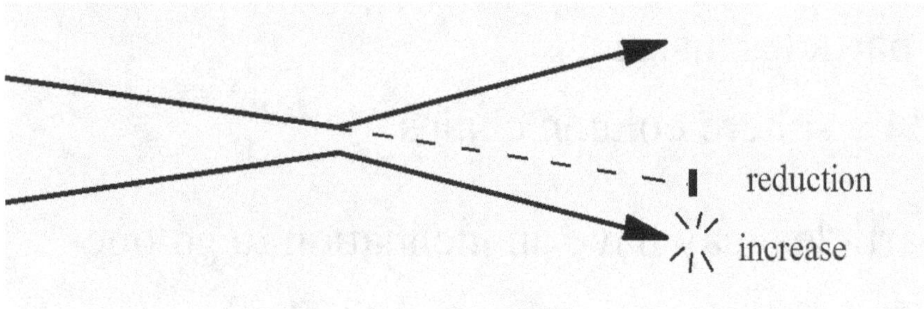
reduction
increase

In the two-slit experiment earlier, some waves deflected each other.

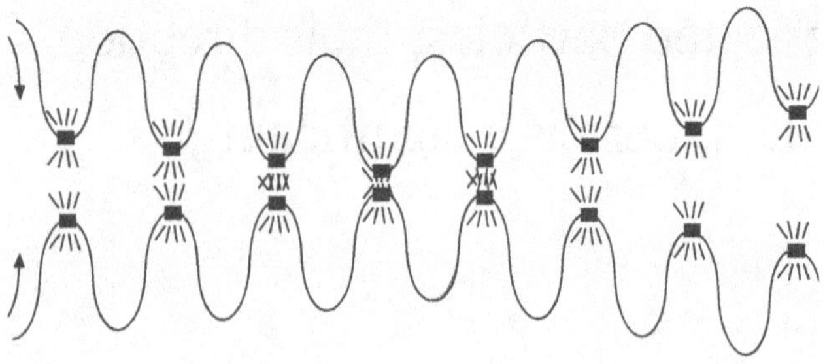

One troughs while the other crests. At more than one location, the waves apply pressure to each other and separate.

The electron version of this experiment also creates an interference pattern. No electrons were destroyed.

A soap bubble is sometimes used to demonstrate wave destruction. One or several colours can appear absent. Some waves pass through, while some are reflected inside the bubble, and others travel along inside the liquid. The result is complicated.

Electromagnetic Wave Structure

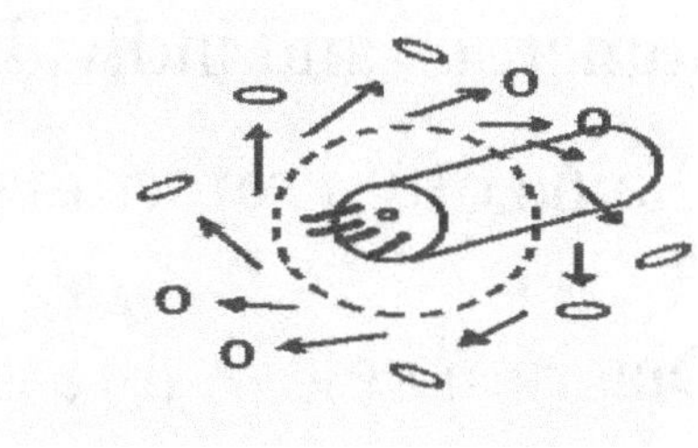

Each flux line particle structure has circling barrels. Barrels, themselves, send out small particle rings.

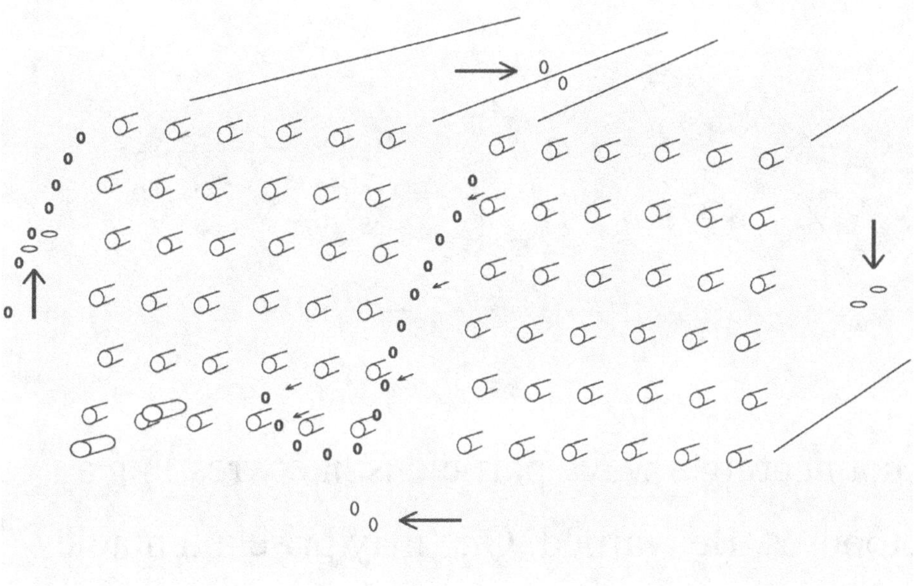

Small particle rings from the side of the wave have space and dominate over the opposite circling direction. These small particle rings are either circling around the wave's flux line particle structures, or they are temporarily curving round the wave and being replaced. There are many of these rings. They all look much the same.

The above diagram represents a portion of a newly formed wave. The wave is travelling into the paper. The nearest end is sending out small particle rings and expanded portions.

Many small particles from the nearest end attract and orientate flux line particles.

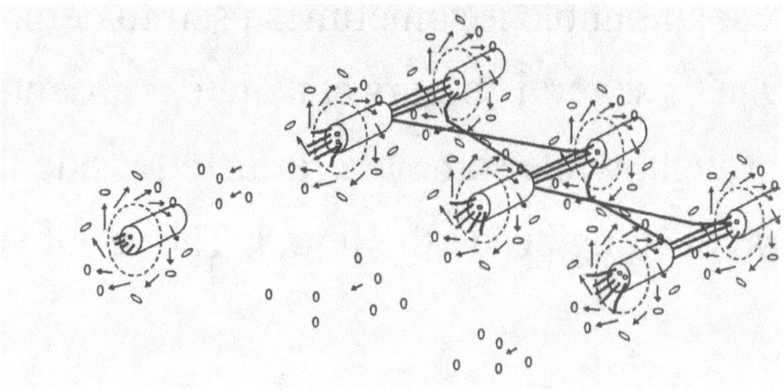

Sphere column chains hold the flux line particle structures in place. These chains are numerous and individually weak. After breaking, sphere column chains often reform very quickly.

Diagram below, a release of small particle rings works its way along a joining sphere column chain.

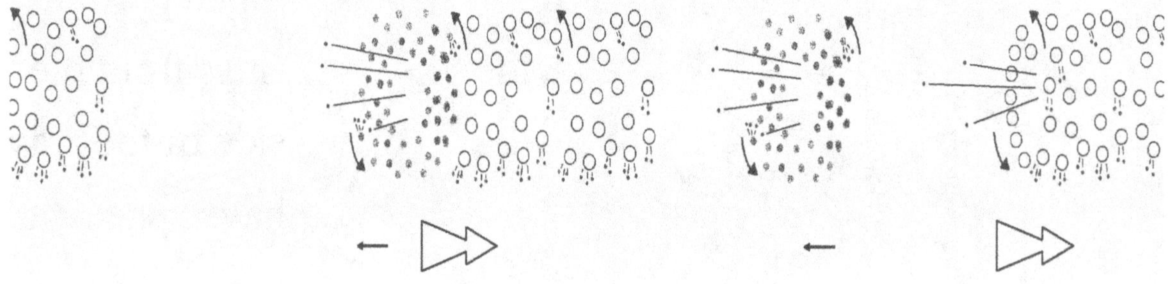

Breaking the joins connecting a wave particle is like wrestling a large number of octopuses. Be warned. One may prise a tentacle off. Many tentacles would remain holding fast.

A large number of joins enables a wave to travel amongst others.

<u>The Speed of Light and Steady Time</u>

By using reflection, the Michelson-Morley experiment divides a beam of light into travelling in two directions, in the direction of Earth's rotation and also at a right-angle to Earth's rotation. Both portions of light are considered to travel at the same speed.

From inside the room of the experiment, the room appears not to move; everything, including the experiment, moves with the rotation of Earth. One light portion is not encountering a larger number of collisions than the other. Collisions may still slow the speed of light.

Light, being particles, has inertia. Light, being reflected back and forth over a long straight line at a right-angle to Earth's turning, would provide confirmation to this idea.

A vacuum is an area without any atoms. Light travels through a vacuum at a speed of 299 792 458 metres per second; this speed is represented by c. Through air the speed has been measured to be approximately 90 000 metres per second slower, and through glass it is usually about two thirds of c.

Finite is able to mean maximum. Light, going from air and then into air of greater density or a cloud, does slow down.

All electromagnetic waves in the same environment travel at approximately the same speed. The global positioning system provides evidence of electromagnetic waves travelling with small adjustments in speed. A little excitingly, the speed adjustments save the need for time to distort. Gravity greatly increases the thickness of the atmosphere around the equator. Arriving from space from different locations, radio waves spend different lengths of time in a varying atmosphere.

An atomic clock relies on a periodic release from atoms of caesium in a magnetic field. These atoms are protected from some bombardment; nonetheless, they don't escape being struck by expanded portions. An atomic clock, after being moved from a tall tower to a position on Earth's surface, finds itself under additional bombardment. This bombardment disrupts the ability of atoms to construct tiny structures including small particle rings. The clock slows while time remains steady.

$$E=mc^2$$

General relativity is a smidgen at odds with this system. The energy equivalence formula stems from the theory of general relativity. The formula, itself, has unlikely tidiness. Each component has units defined long before the formula, defined with no apparent connection nor consideration to the others. A metre could have been a bit longer. A second could have been a bit smaller fraction of a day. Nonetheless, in the formula the units are not adjusted.

Energy(Joules) = mass(Kilograms) x velocity of light squared ((metres per second) squared).

The Universe Does Not Have To Expand To Avoid Becoming One Enormous Mass

An increase in pressure creates heat. A tyre, having been pumped up very hard, provides heat. The heat soon dissipates to an ambient level.

Fossil evidence supports a version of a dragonfly to exist over two hundred million years ago.

A very large constant pressure does not create heat; nonetheless, the centre of Earth remains much hotter than its surface.

We have nearly all material, expanded portions and larger, sending out particles. In the section on gravity particles, we saw them becoming four times more spread out with a doubling of distance. The surface of Earth is sending particles in all directions including inwards. Each time the distance to the centre of Earth is halved, the inwardly moving particles are four times closer together. The material being passed is also sending out particles. The central area, being very busy, is therefore very hot. From bombardment, central material is running low on expanded portions to send out; the attractive reaction is greatly reduced.

While the surface of Earth is attracted towards its sun, mass incorporating the very centre of Earth is hot enough to be repelled.

A mixture of attraction and repulsion also exists between one galaxy and another.

Atomic Structure

In an atom forming environment, antiprotons are bombarded heavily. Being struck from many directions, antiprotons are reduced in number of expanded portions. They move towards protons at a reduced rate. Rather than destruction, strong bonds can be achieved.

Neutrons, protons and very occasionally antiprotons are able to be found after the smashing of a large atom.

Sphere column chains hold wave particles together. Sphere column chains are also able to run from one flux line to another. Flux line particle structures experiencing heavy bombardment lose some circling barrels. A low number of circling barrels helps flux line particles to be alongside each other. A group of strong, compact and secured flux lines is a tethered wave portion.

Without heavy bombardment, neutrons are not known to be greatly attractive to other particles. Inside an atom, neutrons have tethered wave portions at each end. When a neutron comes to be on its own, these wave portions attract and hold onto a disruptive level of material.

The main difference between protons, neutrons and antiprotons is the material that comes with them. Protons, neutrons and antiprotons are approximately the same size. They are main particles.

The Beauty of Tethered Wave Portions

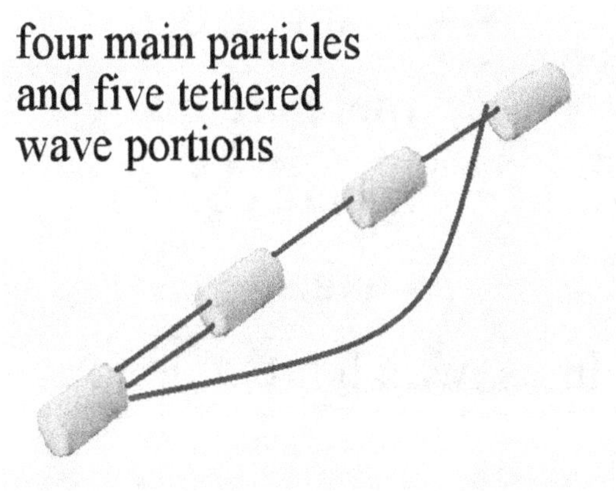
four main particles
and five tethered
wave portions

Many tethered wave portions can be alongside each other. They hold main particles together, and they are under pressure.

Tethered wave portions require a high temperature to form. High temperatures reduce the presence of disrupting material. While the temperature is reducing and nearing the level ceasing all growth of wave portions, disrupting structures are already managing to arrive. Gradual cooling prolongs a period of weaker growth.

Weaker growth occasionally competes with the main structure. To strengthen steel, hot steel can be briefly quenched in water.

Glass is commonly made with a process of gradual cooling. The internal structure has time to adjust without leaving a terrible weakness.

The process of making the very toughest glass makes use of a liquified gas. The gas is used to achieve very rapid cooling from a high temperature; in this way, competing and unhelpful lines are reduced in number.

Structures causing disruption to tethered wave portions are circling barrels of flux line particles.

Tethered wave portions break sometimes.

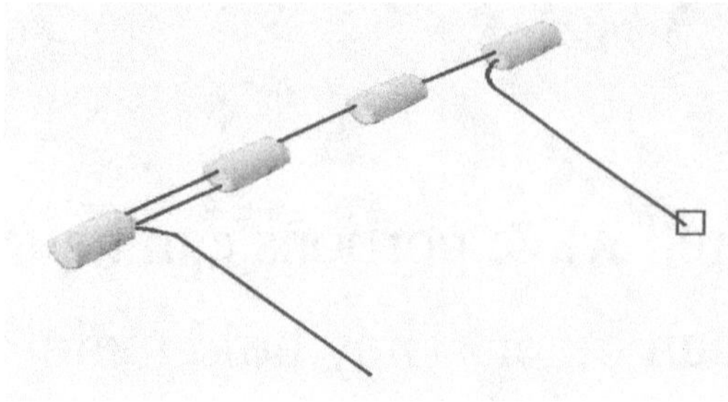

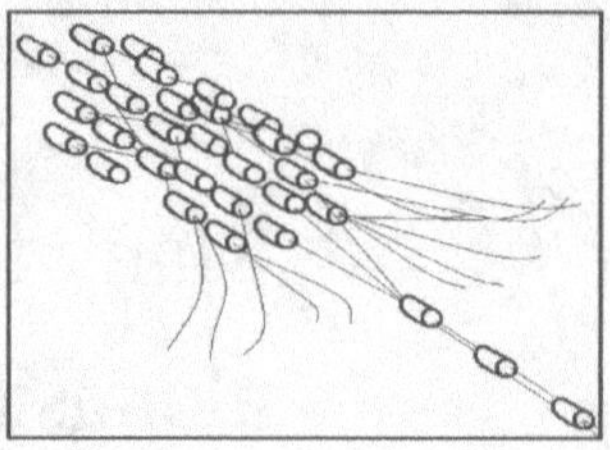

At a free end, there are many sphere column chains.

Frequently, they begin a number of flux lines, which are able to grow.

The space between these flux lines is highly influenced by disruption.

When the distance between flux lines matches an arriving wave, the wave is often blocked. In different areas, a wave has an ability to stretch and compress. This ability helps a wave to be reflected.

Dry glass is an insulator to electricity. Glass is also clear. We can expect glass to have relatively few loose ends compared with many materials. Front cover: a cat enjoys looking at glass; impurities have made the glass brightly coloured.

<u>A Line of Four</u>

When atoms are forming, main particles grow tethered wave portions from their ends. At this time, from heavy bombardment tethered wave portions can reduce in length. One main particle joins with another. They are drawn towards each other until field intensity stops them. Two bonded main particles readily attracts two more. The result is an alpha particle.

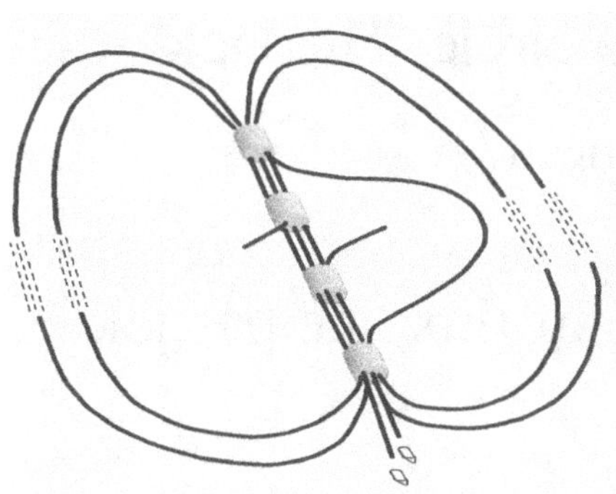

After some calming of the environment, tethered wave portions of each end grow flux lines. The flux lines of one end join with those from the other. Flux lines draw the tethered wave portions towards each other. Bridged by flux lines, a small gap comes to exist between the wave portions. The bond is very secure. This structure is now closed to additional main particles. A solitary line of four gaining two electrons at an end is a single atom of helium.

The Wu experiment supports electrons being at the south end. With the electrons in place, the number of lines leaving each end of the helium atom is approximately equal. Many lines are occupied and stable. The helium atom is unreactive.

With rigorous bombardment, flux lines break. Heavily bombarded atoms of helium are able to join with other atoms.

<u>The Elements</u>

The noble gases of the periodic table have a great reluctance to bond with other atoms. This independence is because of their structures.

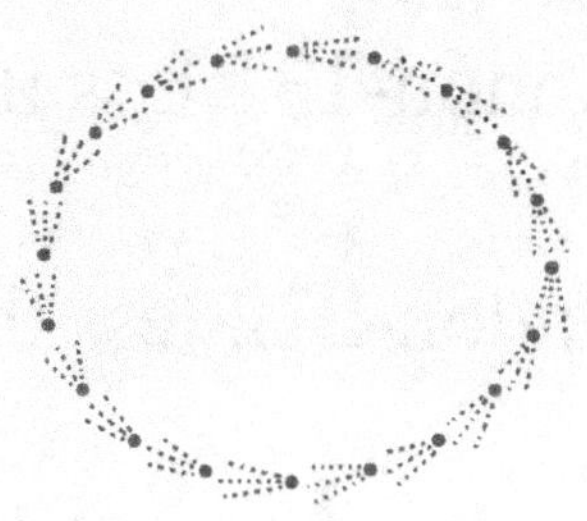

Neon consists of 20 main particles. A circle is a natural shape to consider.

Inside a strong glass container, atoms of neon are put under pressure. An electric current causes bombardment.

Bombardment encourages an atom of neon to circle. The lack of an end to the chain helps to retain the bombardment.

With enough bombardment, there is a release of flux line particle structures in the form of light.

For finding the structures of the remaining inert elements, it is useful to add their protons, neutrons and any antiprotons. The result is the number of main particles, and it is also the atomic mass. Neon has 20 main particles, argon 40, krypton 84, xenon 131, and radon 222.

From neon, argon is the next largest inert element. In the construction of argon, we are able to assume that neon is formed first, and then we are able to find the number of additional particles.

Argon-40 less the structure of neon-20 = 20. Argon may be as neon plus twenty main particles.

Krypton-84 less the structure of argon-40 = 44. Krypton may be as argon plus forty-four main particles.

Xenon-131 less the structure of krypton-84 = 47.

Radon-222 less the structure of xenon-131 = 91.

Expressing each result as a circle leads to the next diagram. A black dot gives an impression of a main particle. Only a few main particles are individually represented.

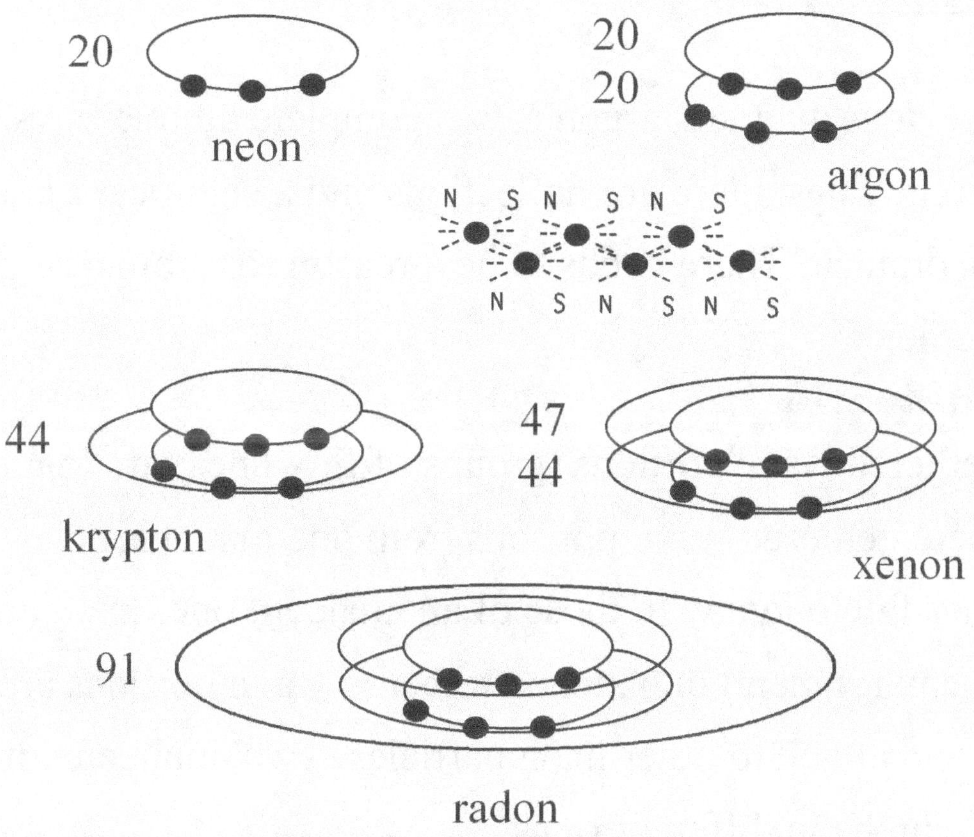
20
neon
20
20
argon
N S N S N S
N S N S N S
44
krypton
47
44
xenon
91
radon

There is a clear pattern, one circle, then one of approximately the same size, a larger circle, then one of approximately the same size, and finally in the diagram, there is an even larger circle.

A reactive element is either a line, or it is a line with a number of circles.

The atomic mass number of a reactive element can be reduced by its completed circles. The remainder forms a line. We now have a method for finding the approximate structure of every naturally occurring atom. There is a small adjustment.

Holding Together

In the previous diagram, argon is two complete atomic rings. Between the rings, there are deflections and a significant level of repelling draught. There needs to be a reason for atomic rings to stay together.

When tethered wave portions (groups of flux lines) are joining with others, tethered wave portions from one main particle sometimes fail to join with those of an adjacent one. In an atom forming environment, unmatched tethered wave portions are free to grow and attach to other main particles. Two connected main particles can be in different rings.

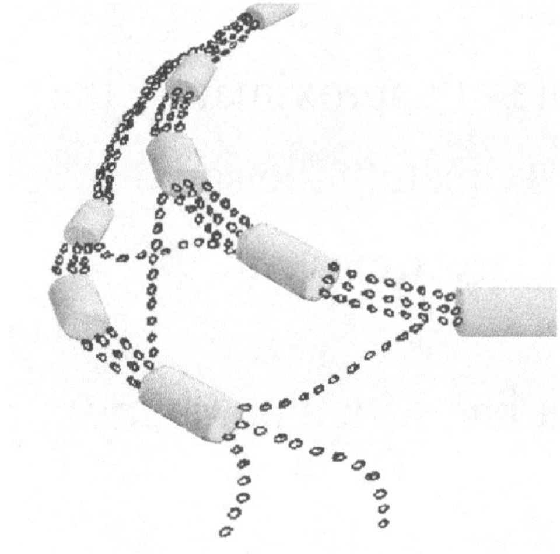

Depicted is part of a ring and a brief outermost line.

Main particles connect with others and form larger structures, until their tethered wave portions are unable to do the task.

Radioactivity

In their clusters, large atoms are under bombardment. Bits of joining material are sent away. Particles at the end of an outermost line become less firmly attached until they, too, are sent away. The field at the distant end of an outermost line is altered by its loss.

N
S

While two electrons are at the end of an outermost line, the field has relatively weak attraction.

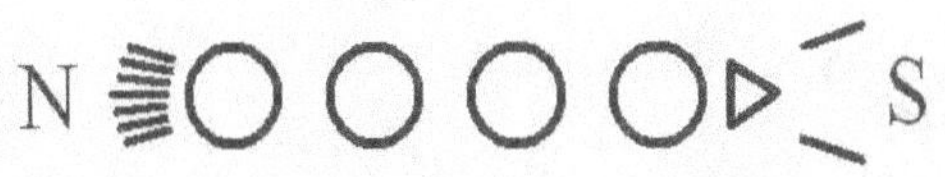
N
S

The loss of an electron causes an increase in field strength.

N
S

The loss of both electrons provides a relatively strong field.

Lead-214 has circles of approximately 16, 16, 44, 51 and an outermost line of 87.

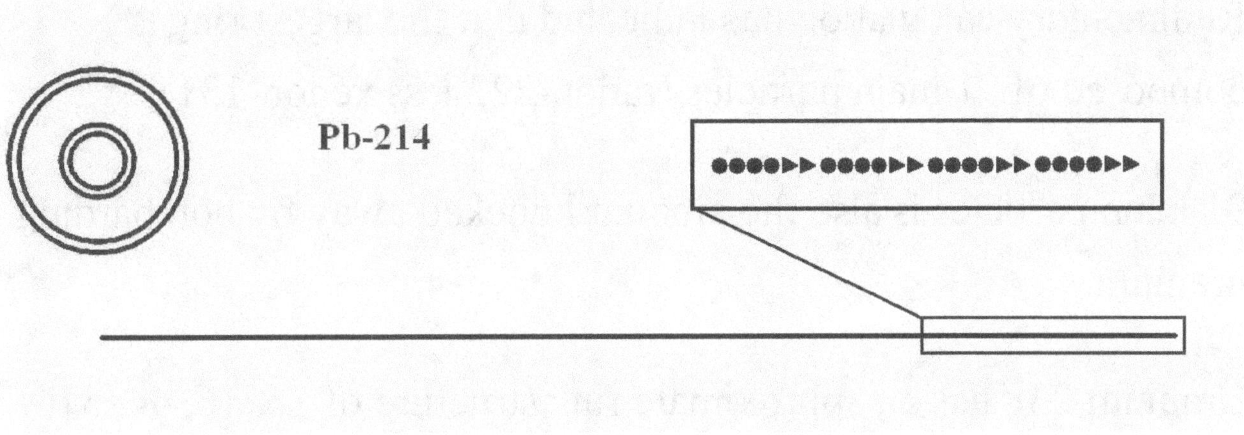
Pb-214

The above represents a single atom of lead-214. Its outermost line would be between two atoms. The upper box contains a representation of the particles inside the lower box. Being lead, this atom has two electrons at its distant end.

After an atom of lead-214 sends out an electron, the change in field causes the atom to be bismuth-214. After another sending out of an electron, the atom is polonium-214. After an atom of polonium-214 sends out an alpha particle, the result is an atom of lead-210.

With a repeat of the same pattern of departure, there is bismuth-210, then polonium-210, and finally lead-206.

Bombardment of Uranium

Rudimentary calculation has indicated that the largest ring is composed of 91 main particles (radon-222 less xenon-131).

91 main particles is also the amount knocked away by bombarding uranium.

Uranium-238 has an approximate ring structure of 16, 16, 44, 51, 91 and an outermost line of 20. The pressure from inner rings provides encouragement for flux lines to be directed outwards. The outermost line of 20 is held by flux lines from inner rings, and the outermost line is also part of a structure involving other atoms.

Fast fission has a neutron fired at a fast rate. When the neutron strikes the outermost and largest ring of an atom of uranium-238, the impact is felt along the entire ring.

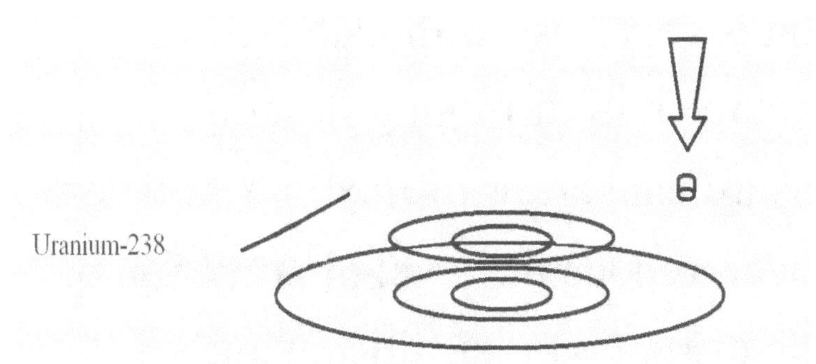
Uranium-238

The outermost ring is knocked away.

With the loss of the ring, promethium occurs with an atomic mass of 147.

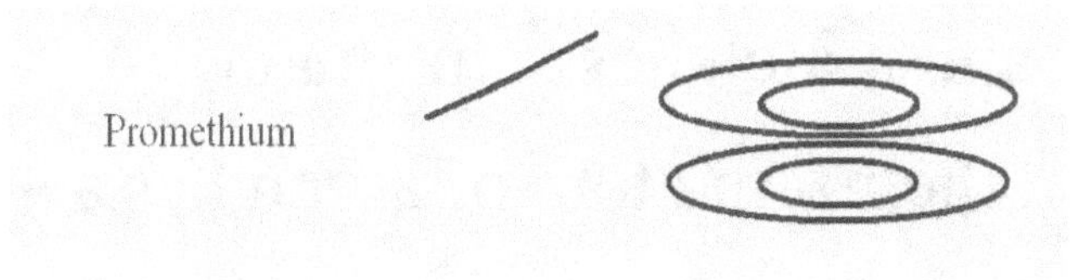
Promethium

Promethium has a ring structure of approximately 16, 16, 44, 51 and an outermost line of 20.

A single main particle leaves from the departing ring of 91. Most likely, this main particle was the first to arrive when the ring of 91 originally began to form. Theoretically, after ninety main particles depart:

one main particle is gripped by the remaining four rings; being no longer a part of the largest atomic ring, the main particle is a target of greatly reduced size; it quickly travels inwards, experiences a large increase in field intensity and is sent away.

The fired neutron and the departing main particle may go on to strike other atoms. They may cause a cascading and dangerous reaction.

<u>The Distribution Table</u>

The following ideas helped to form the table.

An atom forming environment is very busy with material. This intensity discourages the attachment of disrupting structures. Disrupting structures include circling barrels of flux line particles. In large numbers, circling barrels reduce the level of attraction of tethered wave portions. While disruption is significant and increasing, main particles attach at an increasing distance until they are incapable of joining.

An atom of neon is one ring of 20, and it includes some main particles with added space between them. An atom of argon has a structure of 16, 24. Disruption began to grow before the second ring has 16 main particles. With less attraction, more main particles are required to complete the second ring.

Atoms of group 2 of the periodic table have a longer outermost line than those of group 1. Potassium 16, 19, 4 of group 1 has some less attracting particles in the second ring and in its outermost line. Calcium 16, 16, 8 of group 2 has them mainly in its outermost line.

In the formation of fourth and fifth rings, a growing outermost line flaps wildly about. Outward pressure and long lines are demanding on bonds. Fourth and fifth rings either achieve a complete ring with little disruption, or they fail to complete. If a fourth ring completes and is with no further growth, the ring requires 55 main particles. 51 main particles is reserved for easy completion. With the same idea, fifth rings use 95 to 91 main particles.

The table may only be a stepping stone towards even greater accuracy.

Element	Atomic Weight	Mass	Rough Distribution of Main Particles	Symbol	Reference Number
Actinium	227	227	16 16 44 51 91 9	Ac	89
Aluminium	26.98	27	16 11	Al	13
Americium	243	243	16 16 44 51 91 25	Am	95
Antimony	121.7	122	16 16 44 46	Sb	51
Argon	39.94	40	16 24	Ar	18
Arsenic	74.92	75	16 16 43	As	33
Astatine	210	210	16 16 44 51 83	At	85
Barium	137.34	137	16 16 44 53 8	Ba	56
Berkelium	247	247	16 16 44 51 91 29	Bk	97
Beryllium	9.01	9	9	Be	4
Bismuth	208.98	209	16 16 44 51 82	Bi	83
Boron	10.81	11	11	B	5
Bromine	79.9	80	16 16 48	Br	35
Cadmium	112.4	112	16 16 44 36	Cd	48
Caesium	132.91	133	16 16 44 53 4	Cs	55
Calcium	40.08	40	16 16 8	Ca	20
Californium	251	251	16 16 44 51 91 33	Cf	98
Carbon	12.01	12	12	C	6
Cerium	140.12	140	16 16 44 51 13	Ce	58
Chlorine	35.45	35	16 19	Cl	17
Chromium	52	52	16 16 20	Cr	24
Cobalt	58.93	59	16 16 27	Co	27
Copper	63.54	64	16 16 32	Cu	29
Curium	247	247	16 16 44 51 91 29	Cm	96
Dysprosium	162.5	163	16 16 44 51 36	Dy	66
Einsteinium	254	254	16 16 44 51 91 36	Es	99
Erbium	167.2	167	16 16 44 51 40	Er	68
Europium	151.96	152	16 16 44 51 25	Eu	63
Fermium	257	257	16 16 44 51 91 39	Fm	100
Fluorine	19	19	19	F	9
Francium	223	223	16 16 44 51 91 5	Fr	87
Gadolinium	157.2	157	16 16 44 51 30	Gd	64
Gallium	69.72	70	16 16 38	Ga	31
Germanium	72.5	73	16 16 41	Ge	32
Gold	196.97	197	16 16 44 51 70	Au	79
Hafnium	178.4	178	16 16 44 51 51	Hf	72
Hahnium	268	268	16 16 44 51 91 50	Ha	105
Helium	4	4	4	He	2
Holmium	164.93	165	16 16 44 51 38	Ho	67
Hydrogen	1.01	1	1	H	1
Indium	114.82	115	16 16 44 39	In	49
Iodine	126.9	127	16 16 44 51 NCR	I	53
Iridium	192.2	192	16 16 44 51 65	Ir	77
Iron	55.84	56	16 16 24	Fe	26
Krypton	83.8	84	16 16 52	Kr	36
Lanthanum	138.91	139	16 16 44 51 12	La	57

Element	Atomic Weight	Atomic Mass	Rough Distribution of Main Particles	Symbol	Reference Number
Lawrencium	256	256	16 16 44 51 91 38	Lr	103
Lead	207.2	207	16 16 44 51 80	Pb	82
Lithium	6.94	7	7	Li	3
Lutetium	174.97	175	16 16 44 51 48	Lu	71
Magnesium	24.31	24	16 8	Mg	12
Manganese	54.94	55	16 16 23	Mn	25
Mendelevium	257	257	16 16 44 51 91 39	Md	101
Mercury	200.5	201	16 16 44 51 74	Hg	80
Molybdenum	95.9	96	16 16 44 20	Mo	42
Neodymium	144.2	144	16 16 44 51 17	Nd	60
Neon	20.17	20	20	Ne	10
Neptunium	237.05	237	16 16 44 51 91 19	Np	93
Nickel	58.7	59	16 16 27	Ni	28
Niobium	92.91	93	16 16 44 17	Nb	41
Nitrogen	14.01	14	14	N	7
Nobelium	255	255	16 16 44 51 91 37	No	102
Osmium	190.2	190	16 16 44 51 63	Os	76
Oxygen	16	16	16	O	8
Palladium	106.4	106	16 16 44 30	Pd	46
Phosphorus	30.97	31	16 15	P	15
Platinum	195	195	16 16 44 51 68	Pt	78
Plutonium	244	244	16 16 44 51 91 26	Pu	94
Polonium	209	209	16 16 44 51 82	Po	84
Potassium	39.09	39	16 19 4	K	19
Praseodymium	140.91	141	16 16 44 51 14	Pr	59
Promethium	147	147	16 16 44 51 20	Pm	61
Protactinium	231.04	231	16 16 44 51 91 13	Pa	91
Radium	226.03	226	16 16 44 51 91 8	Ra	88
Radon	222	222	16 16 44 51 95	Rn	86
Rhenium	186.2	186	16 16 44 51 59	Re	75
Rhodium	102.91	103	16 16 44 27	Rh	45
Rubidium	85.47	85	16 16 49 4	Rb	37
Ruthenium	101	101	16 16 44 25	Ru	44
Rutherfordium	267	267	16 16 44 51 91 49	Rf	104
Samarium	150.4	150	16 16 44 51 23	Sm	62
Scandium	44.96	45	16 16 13	Sc	21
Selenium	78.9	79	16 16 47	Se	34
Silicon	28.08	28	16 12	Si	14
Silver	107.87	108	16 16 44 32	Ag	47
Sodium	22.99	23	19 4	Na	11

Element	Atomic		Rough Distribution of		Reference
	Weight	Mass	Main Particles	Symbol	Number
Strontium	87.62	88	16 16 48 8	Sr	38
Sulphur	32.06	32	16 16	S	16
Tantalum	180.95	181	16 16 44 51 54	Ta	73
Technetium	97	97	16 16 44 21	Tc	43
Tellurium	127.6	128	16 16 44 52 NCR	Te	52
Terbium	158.93	159	16 16 44 51 32	Tb	65
Thallium	204.3	204	16 16 44 51 77	Tl	81
Thorium	232.04	232	16 16 44 51 91 14	Th	90
Thulium	168.93	169	16 16 44 51 42	Tm	69
Tin	118.6	119	16 16 44 43	Sn	50
Titanium	47.9	48	16 16 16	Ti	22
Tungsten	183.8	184	16 16 44 51 57	W	74
Uranium	238.03	238	16 16 44 51 91 20	U	92
Vanadium	50.94	51	16 16 19	V	23
Xenon	131.3	131	16 16 44 55	Xe	54
Ytterbium	173	173	16 16 44 51 46	Yb	70
Yttrium	88.91	89	16 16 44 13	Y	39
Zinc	65.38	65	16 16 33	Zn	30
Zirconium	91.22	91	16 16 44 15	Zr	40

Small variations in number of main particles inside a ring are insignificant.

In the column for the ring structure of tellurium, there is NCR, not a complete ring. The number of main particles is large enough to complete the fourth ring; nonetheless, disruption causes a gap suitable for bonding. Tellurium is able to bond with atoms of group 2.

Iodine bonds with atoms of group 1. From being less disrupted than tellurium, an atom of iodine is a tighter structure, and it possesses a smaller gap than the gap in a tellurium atom.

The Periodic Table

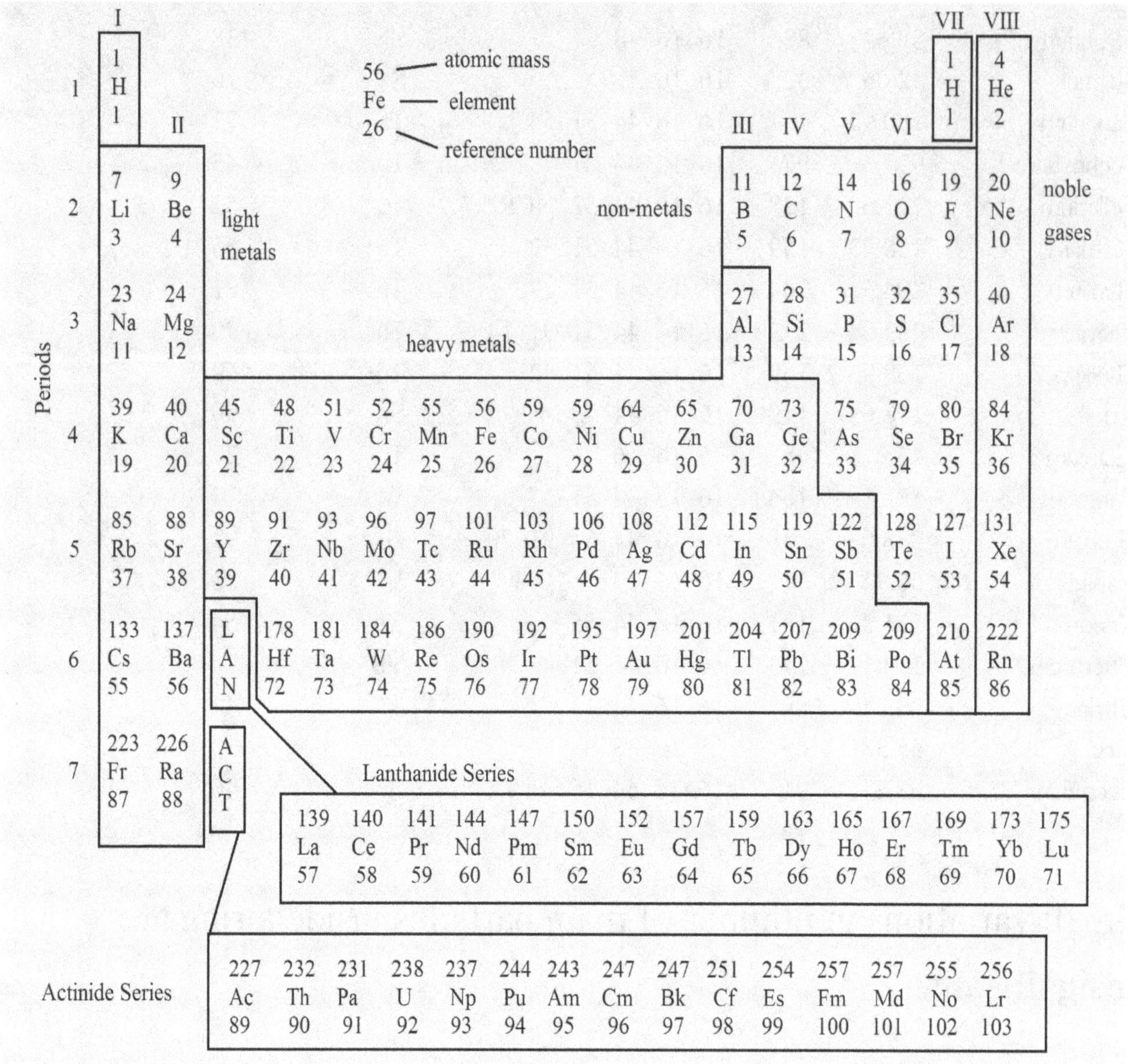
I
II
III IV V VI VII VIII
56 atomic mass
Fe element
26 reference number
Periods
1 H 1
1 H 1
4 He 2
7 Li 3
9 Be 4
light metals
non-metals
11 B 5
12 C 6
14 N 7
16 O 8
19 F 9
20 Ne 10
noble gases
23 Na 11
24 Mg 12
heavy metals
27 Al 13
28 Si 14
31 P 15
32 S 16
35 Cl 17
40 Ar 18
39 K 19
40 Ca 20
45 Sc 21
48 Ti 22
51 V 23
52 Cr 24
55 Mn 25
56 Fe 26
59 Co 27
59 Ni 28
64 Cu 29
65 Zn 30
70 Ga 31
73 Ge 32
75 As 33
79 Se 34
80 Br 35
84 Kr 36
85 Rb 37
88 Sr 38
89 Y 39
91 Zr 40
93 Nb 41
96 Mo 42
97 Tc 43
101 Ru 44
103 Rh 45
106 Pd 46
108 Ag 47
112 Cd 48
115 In 49
119 Sn 50
122 Sb 51
128 Te 52
127 I 53
131 Xe 54
133 Cs 55
137 Ba 56
L A N
178 Hf 72
181 Ta 73
184 W 74
186 Re 75
190 Os 76
192 Ir 77
195 Pt 78
197 Au 79
201 Hg 80
204 Tl 81
207 Pb 82
209 Bi 83
209 Po 84
210 At 85
222 Rn 86
223 Fr 87
226 Ra 88
A C T
Lanthanide Series
139 La 57
140 Ce 58
141 Pr 59
144 Nd 60
147 Pm 61
150 Sm 62
152 Eu 63
157 Gd 64
159 Tb 65
163 Dy 66
165 Ho 67
167 Er 68
169 Tm 69
173 Yb 70
175 Lu 71
Actinide Series
227 Ac 89
232 Th 90
231 Pa 91
238 U 92
237 Np 93
244 Pu 94
243 Am 95
247 Cm 96
247 Bk 97
251 Cf 98
254 Es 99
257 Fm 100
257 Md 101
255 No 102
256 Lr 103

Periods are horizontal. Moving from left and towards the right along a period, the outermost line possessed by each atom is increasingly close to being able to complete a circle. There are very occasional exceptions. Hydrogen, busy enjoying two locations, is better placed in group 1. Cobalt and nickel have the same distribution of main particles. Cobalt has an additional closely held electron at the end of its outermost line.

Groups are vertical. Elements of a group have similar traits, because they have a similarity in structure.

Atoms of an element of group 1 have brief outermost lines. When they cluster, tight structures form. Any near atomic rings add to the pressure. Elements of group 1 are therefore soft. They have low melting points, and they react readily with suitable substances. An atom of lithium has no disruptive atomic rings, and it is longer in length than hydrogen. At 180° Celsius, lithium has the highest melting point of group 1.

Elements of group 1 fit with group 7, and elements of group 2 fit with group 6. One group provides a filling, and the other group provides a gap.

Rather than fill or form gaps, atoms of a heavy metal are more ready to cluster. The field of the circles encourages the outermost line to be unwrapped. In diagram, one heavy metal atom joins with another.

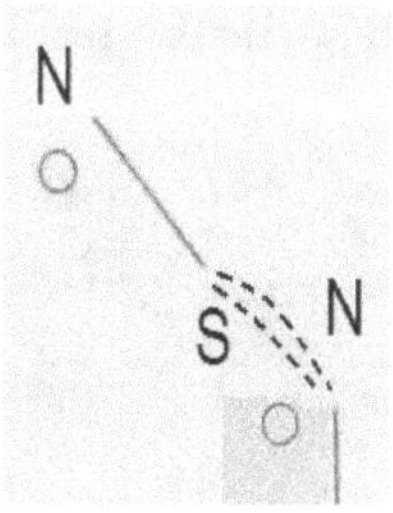
N
O
S
N
O

An atom of silica has a structure of 16, 12. Its outermost line is able to curve round and attract itself. By itself, it is able to form a gap, and it is a non-metal. Oxygen is a non-metal without an atomic ring. Two atoms of oxygen readily bond to each other rather than cluster as a metal.

Group 8 is sometimes sensibly called group 0. Each one is a number of complete atomic rings, except helium. Helium, although normally unreactive, has different traits to the rest of its group. Helium does not produce light as readily as the other members. Atoms of helium can be aligned, and they make a wonderfully powerful magnet. These magnets are used in Earth's largest particle accelerator.

Electricity

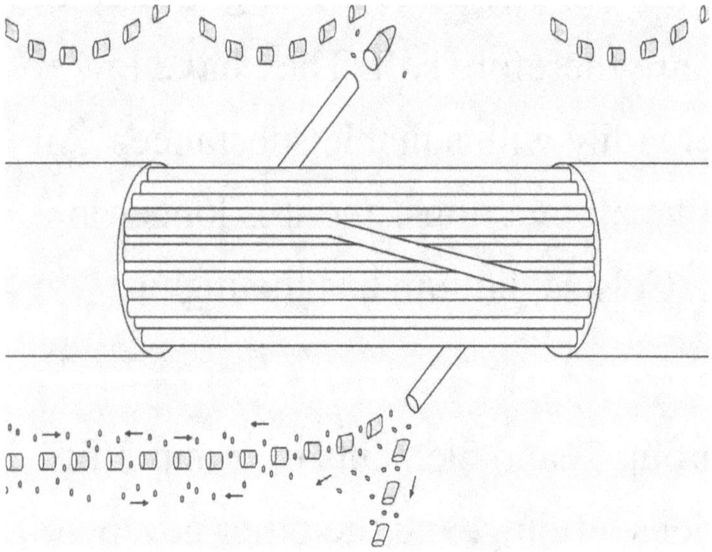

Upper middle, the south end of a broken tethered wave portion holds an electron.

Low in diagram, flux lines are held by the north end of a broken tethered wave portion.

To scale, flux lines and small particle rings would be smaller in size and larger in number.

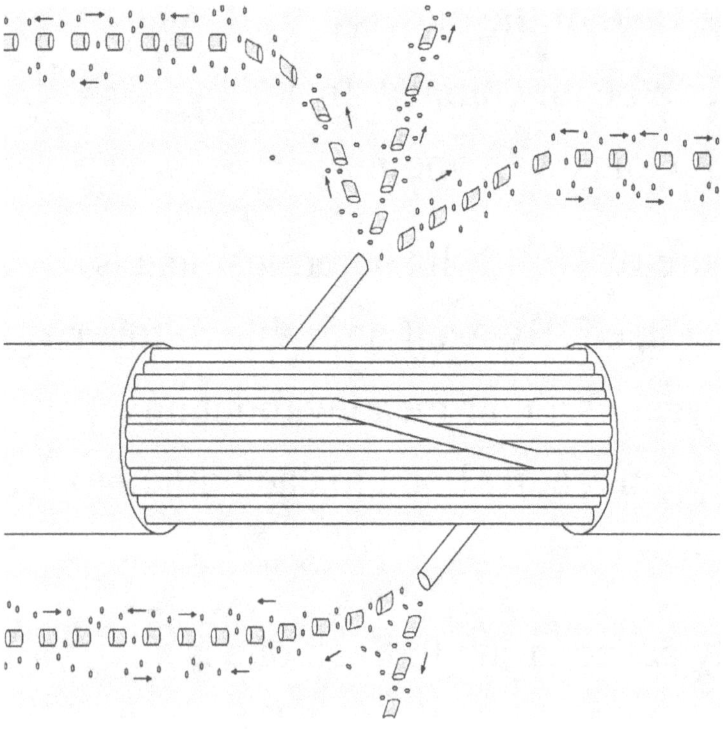

The electron has been extracted. Flux lines occupy its position.

The absence of many electrons at one end of a wire causes some alignment of flux lines in the wire. Alignment travels along a copper wire and slowly reduces.

Small particle rings from the flat ends of flux line particles tend to be more developed than those from the sides. From the ends, south leading small particle rings travel in one direction, and north leading rings travel in the opposite direction.

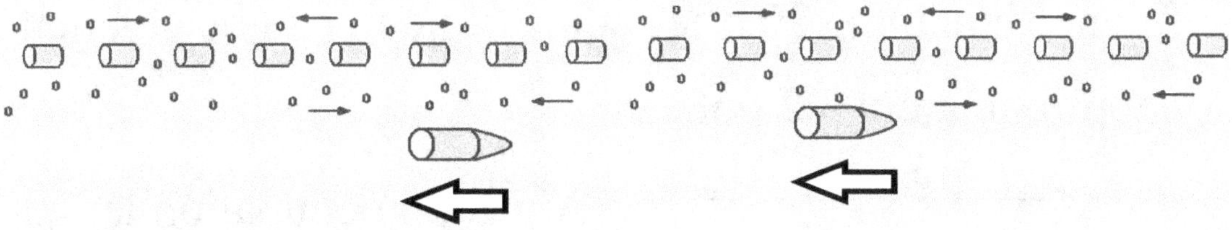

Above, many south leading small particle rings travel to the right. Electrons have been orientated, and they are attracted to the left.

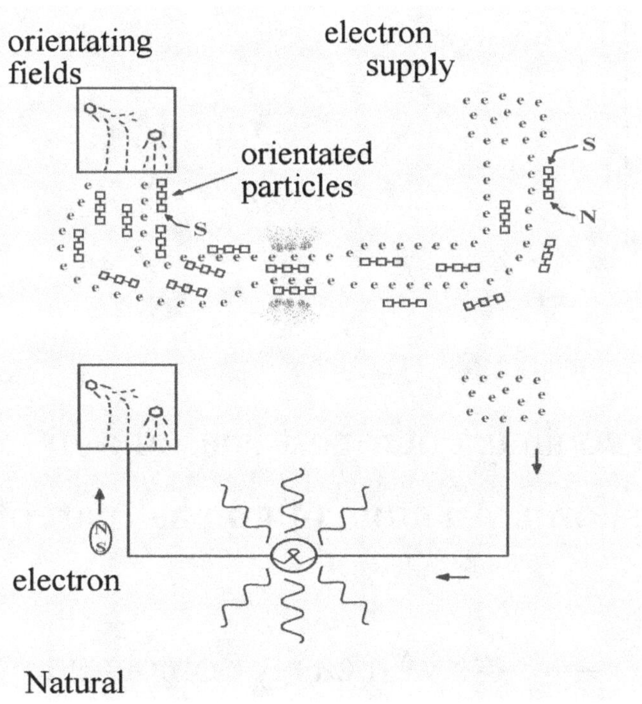

orientating fields
electron supply
orientated particles
S
N
electron
Natural

Electrons follow the path of increasing alignment.

While electrons travel along a wire, they bombard each other and the wire with expanded portions, small particle rings and flux line particle structures.

Due to bombardment, flux line particle structures are released from main particles and electrons. The flux line particle structures form electromagnetic waves.

Relative Ring Sizes and the Proton Occupation Length

A neon atom involves 20 main particles. Each main particle can be considered to occupy a length. This length can be called a proton occupation length or pol.

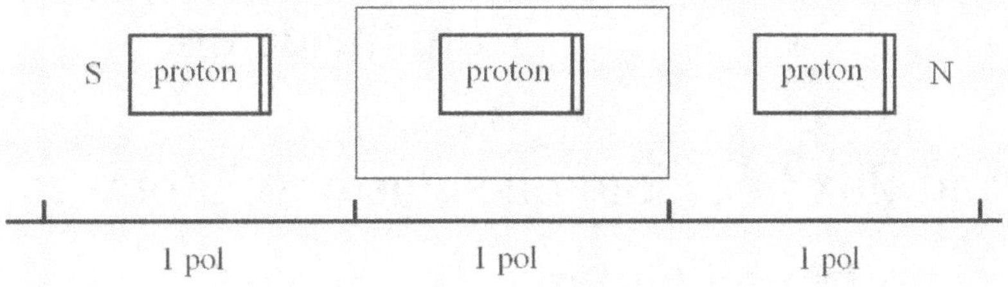

S
proton
proton
proton
N
1 pol
1 pol
1 pol

The circumference of neon is then about 20 pol. Using $C=2\pi r$ and circumference equalling 20, the radius of neon can be calculated to be very approximately 3.2 pol.

Circumference	Radius
16	2.5
20	3.2
44	7
51	8.1
55	8.75
91	14.5
95	15.1

A proton occupation length is a very approximate unit of relative length. It is useful in the drawing of atoms.

Iron and Magnetic Structure

Atoms of iron are created. While cooling, outermost lines of iron atoms attract each other. Clusters form. An atom of iron has part of a third ring.

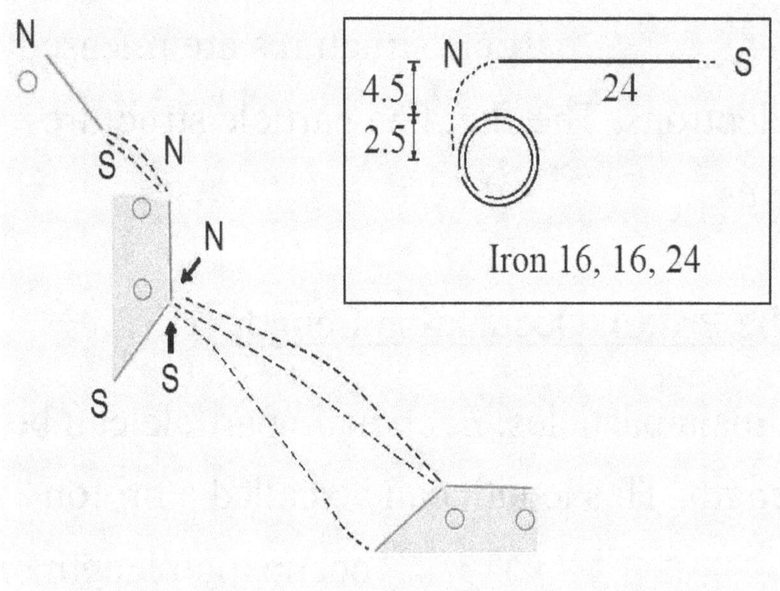
N
S
N
N
S
S
4.5
2.5
N
24
S
Iron 16, 16, 24

An easily completed third ring of an atom consists of 44 main particles. Assuming a line begins at the same distance as a circle, the north end of this outermost line is 7 pol from the centre.

Tethered wave portions and flux lines join one cluster of atoms to another. The next diagram portrays a pattern.

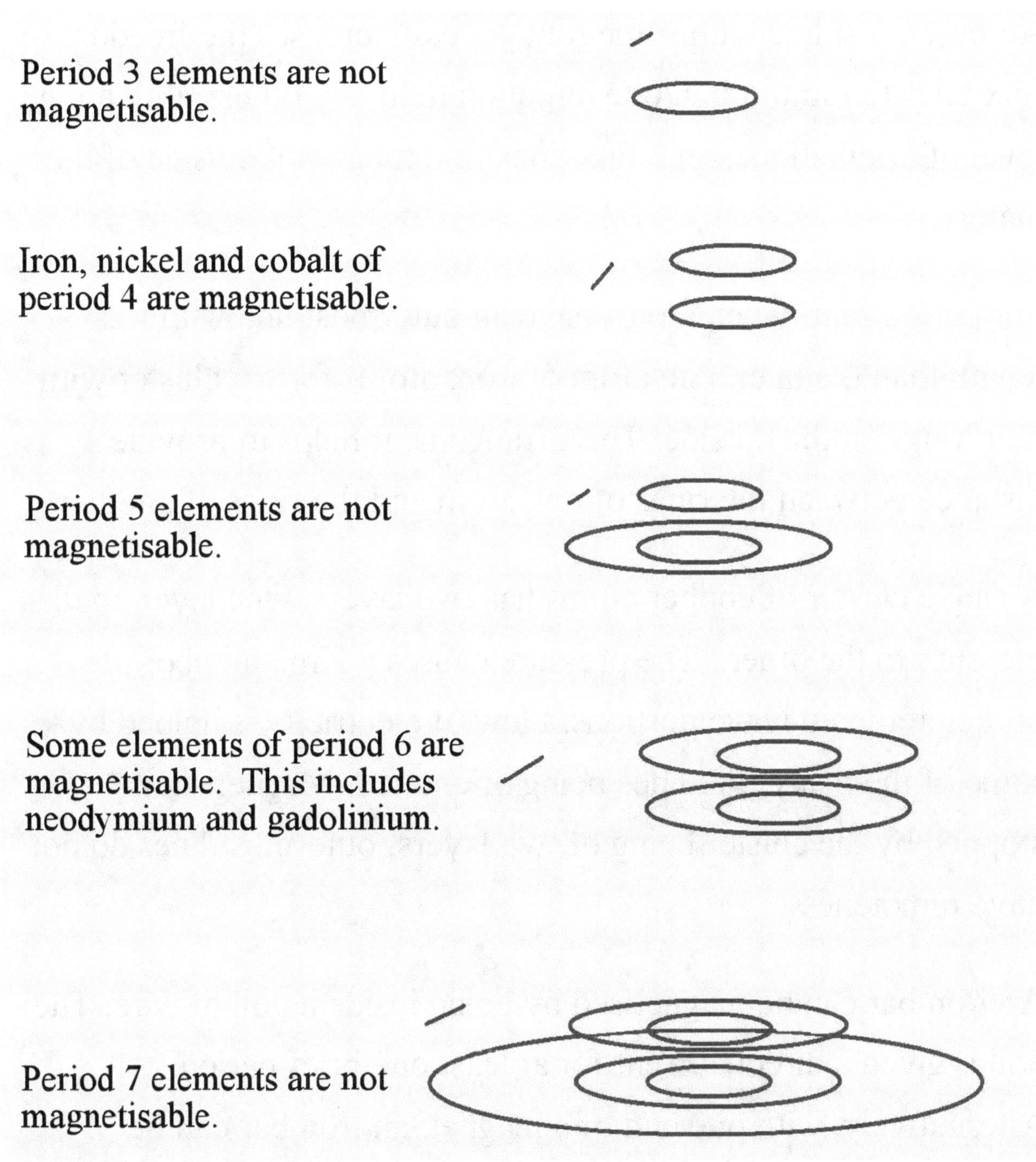
Period 3 elements are not magnetisable.
Iron, nickel and cobalt of period 4 are magnetisable.
Period 5 elements are not magnetisable.
Some elements of period 6 are magnetisable. This includes neodymium and gadolinium.
Period 7 elements are not magnetisable.

An element suitable for magnetisation must have an outermost line with some remoteness.

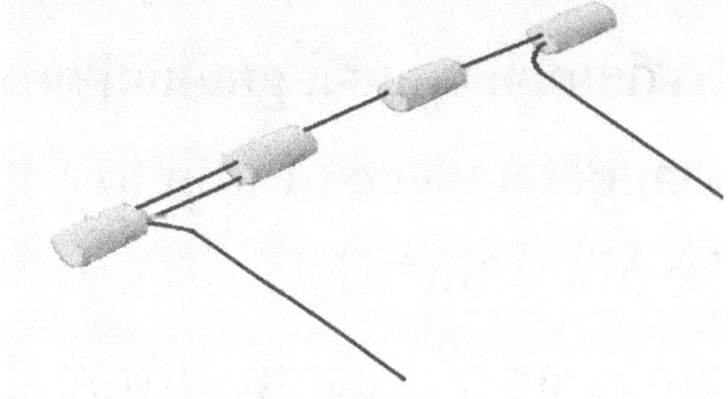

Under strain, tethered wave portions break sometimes. This occurrence creates two unattached ends, one with north polarity and one with south. An unattached end branches out into a number of flux lines.

An outermost line with some remoteness increases the freedom of flux lines to run more in one direction than any other. With a high level of orientation, some flux lines find their way outside of the material.

For period four, when atoms have an outermost line with less length than the line of an atom of iron, atoms form a cluster with their rings on the outside. This arrangement helps to provide distance between the rings of one atom and the rings of another.

A single cluster of copper atoms has two layers. One layer applies pressure to the other. The pressure causes a large number of broken tethered wave portions. Flow of electricity is helped by some of the many flux lines being orientated. Magnetism is stopped by the cluster being of two layers; outermost lines do not have remoteness.

An iron bar can be magnetised by being inside a coil of wire. The coil is given a direct current for at least one brief period. Alternatively, with one end of a magnet, an iron bar can be repeatedly stroked in one direction. Both methods cause some alignment.

To demonstrate a gradual breaking and adjustment of flux lines, connect a piece of iron to a magnet. At least for a length of time, the iron slowly increases in magnetisation. The iron can be temporarily separated from the magnet, and its own magnetic strength can be measured.

Magnetic Field of an Electric Current

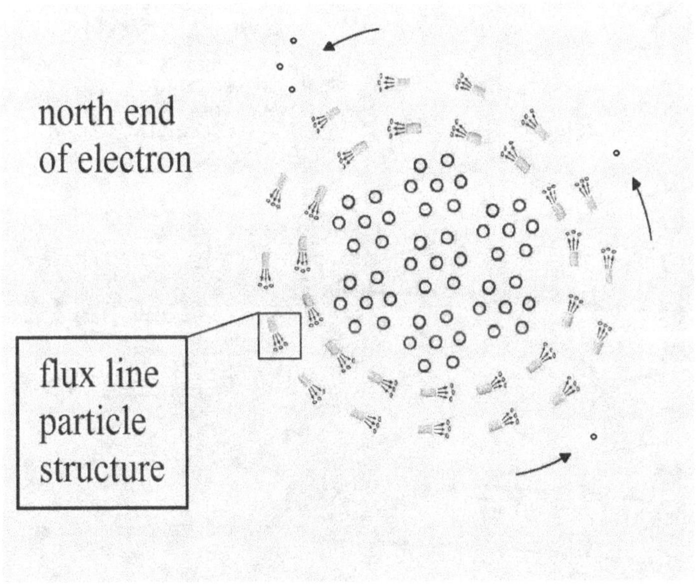

north end
of electron
flux line
particle
structure

At the north end of electrons, flux line particle structures are circling clockwise. South leading small particle rings are sent out in a roughly counterclockwise direction. The side of the small particle rings closest to the electron encounters more material including expanded portions. Encountering more material on the one side causes the small particle rings to include a curving path. In the diagram above, the arrows express the movement of the small particle rings.

Small particle rings are able to orientate and attract flux line particles.

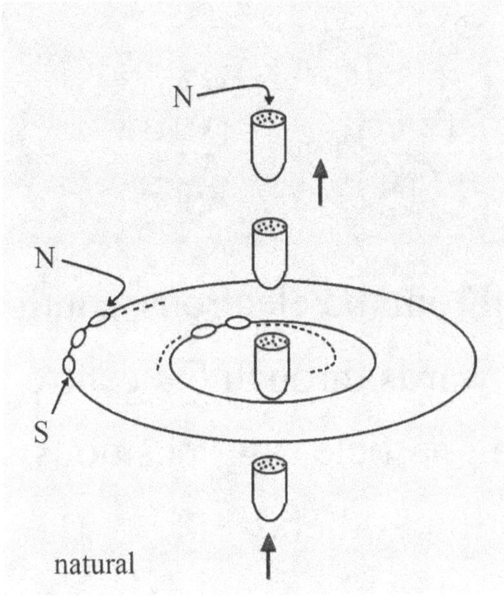

N
N
S
natural

In the diagram, the north face is indicated.

While electrons travel along a wire, flux line particles are attracted into making circles.

The flux line particles are orientated as in the large circle.

Electromagnetism

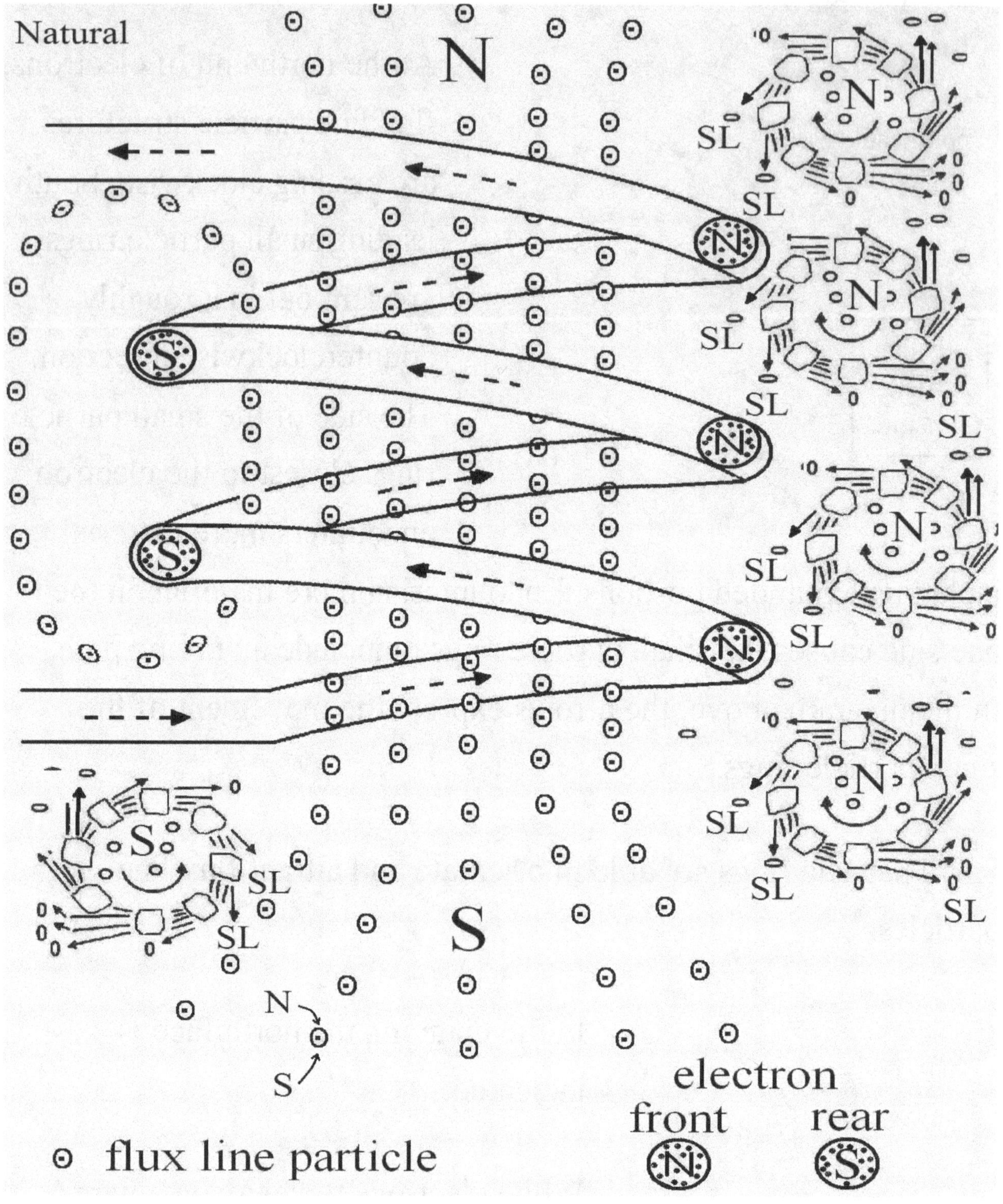
Natural
N
S
SL
N
S
electron
front
rear
flux line particle

Electrons of a current run along the coil. From the electrons, south leading small particle rings are sent downwards through the centre of the coil. The concentrated flow of small particle rings positions flux line particles. Established particles attract others.

Looking towards an end of a coil, electrons travelling clockwise indicates north polarity. Electrons travelling counterclockwise indicates south polarity.

Heat

Small moving particles cause disruption and encourage rearrangement. These particles are heat.

An exothermic chemical reaction creates a structure which overall causes itself greater bombardment. The structure sends particles into the environment.

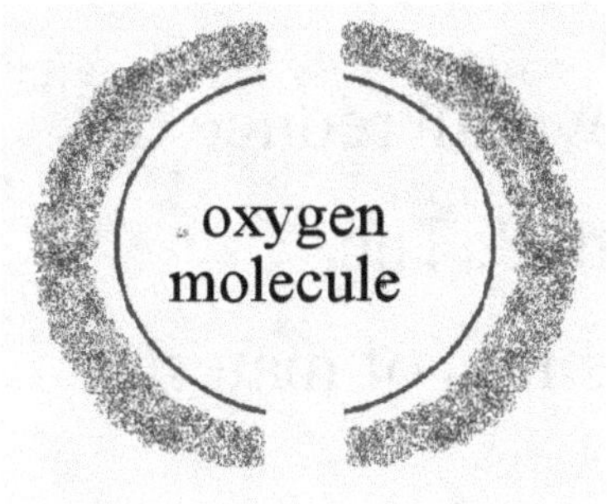
oxygen
molecule

Combining oxygen gas, hydrogen gas and heat creates an explosive exothermic reaction.

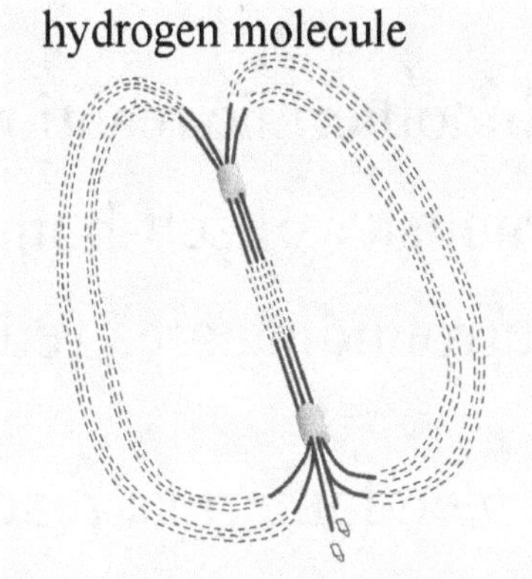
hydrogen molecule

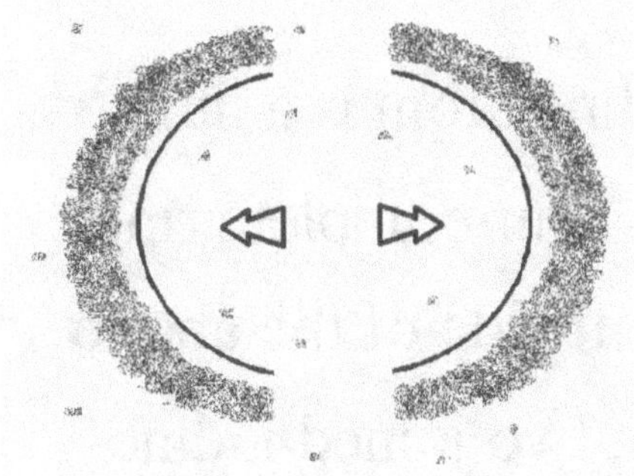

With added heat, one atom bombards the other an increased amount.

In the disruptive clashes, flux lines break. Flux lines grow, and they attach.

A pair of hydrogen atoms is a much smaller target than an atom of oxygen. The pair moves in, and it replaces one atom of oxygen.

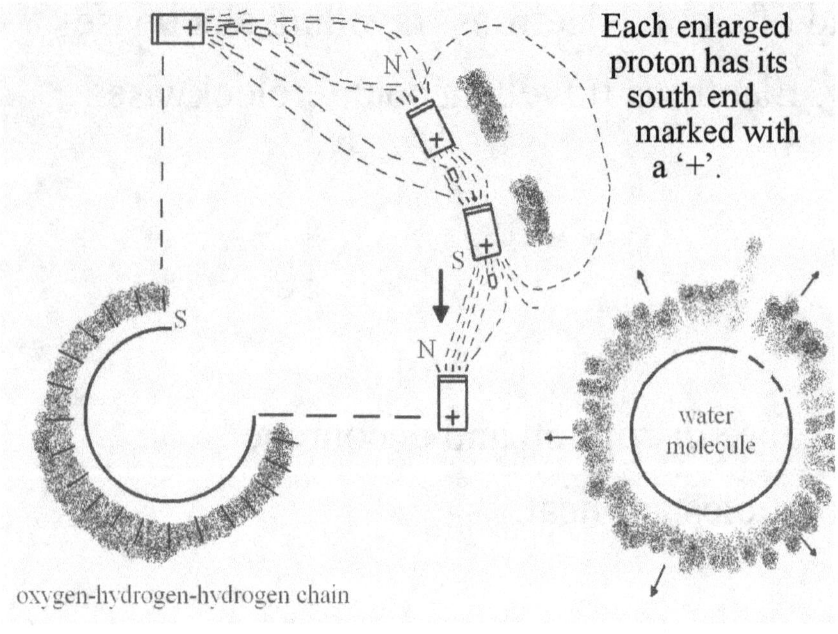
S
N
S
N
S
water
molecule
Each enlarged proton has its south end marked with a '+'.
oxygen-hydrogen-hydrogen chain

With flux lines drawing in the atoms, the result is a tighter structure and further bombardment. The environment is warmed by additional moving material.

Three molecules have turned into two molecules of water. If each molecule had small surrounding particles, then the lower number of molecules most likely added to the heat output.

Endothermic reactions create structures which overall reduce the intensity of self-bombardment. Growth is permitted. The surrounding area reduces in temperature from its loss of material.

An endothermic reaction is $H_2 + I_2 \rightarrow 2HI$.

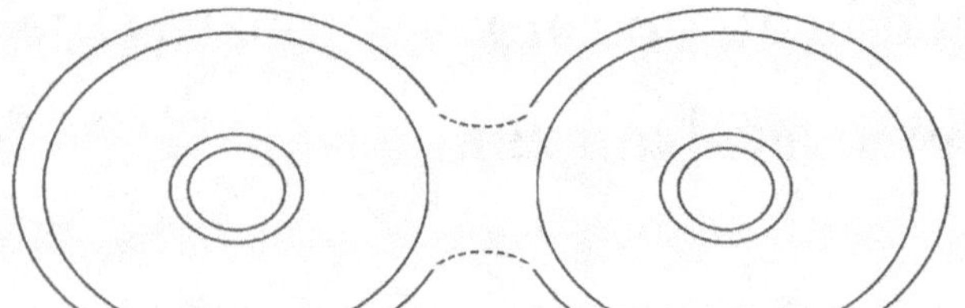

Each iodine atom is a little less than four complete rings. On the left side of the above equation, two joined iodine atoms bombard each other. On the right side of the equation, the iodine atoms have separated. Each one has joined with a hydrogen atom. Overall, the new structures cause themselves less bombardment, and they gain flux line particle structures from the environment.

Solids, Liquids and Gases

If not all, most solids have clusters of atoms. A cluster of a solid allows some flux lines to join other clusters in a stable manner.

Liquids are atoms or molecules which are drawn in, but they are not fixed in one area. Flux lines joining one group to another are continually breaking and reforming.

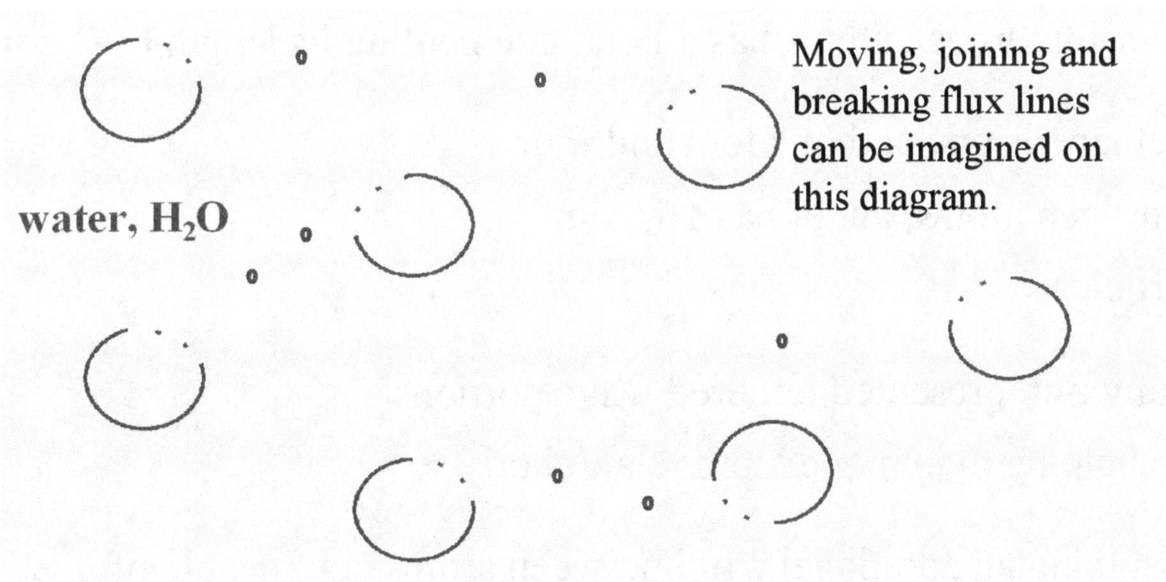
water, H_2O
Moving, joining and breaking flux lines can be imagined on this diagram.

A gas particle can be an atom or a molecule. A gas particle comprises of a closed shape. An atom of any member of group 8 is a closed shape of main particles except atoms of helium. An atom of helium with stable flux lines can be considered to be a closed shape, diagram in ‘A Line of Four’. At most, flux lines are weak between gas particles.

Two atoms of oxygen form a closed shape involving two very separate bonds.

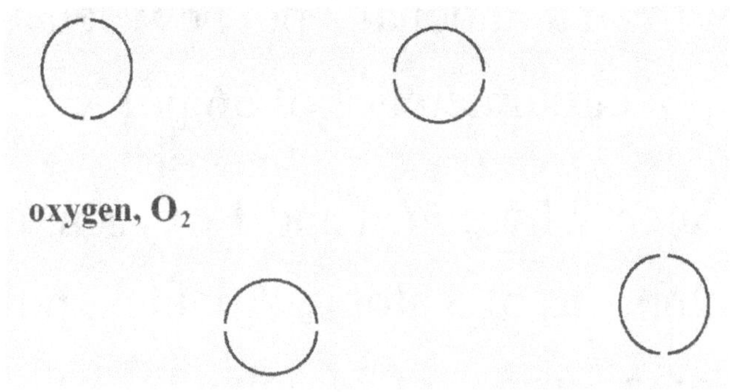
oxygen, O_2

Molecular Bonding

Compounds are formed in a calmer environment than an atom forming environment. Bonds are less stressed, and a weaker level of bond becomes possible.

At room temperature, an atom of helium is unreactive. A pair of hydrogen atoms is a gas particle. With the idea that the stability of these structures is a result of them having a similar length, then a pair of hydrogen atoms has a bond augmenting its length by 2 pol.

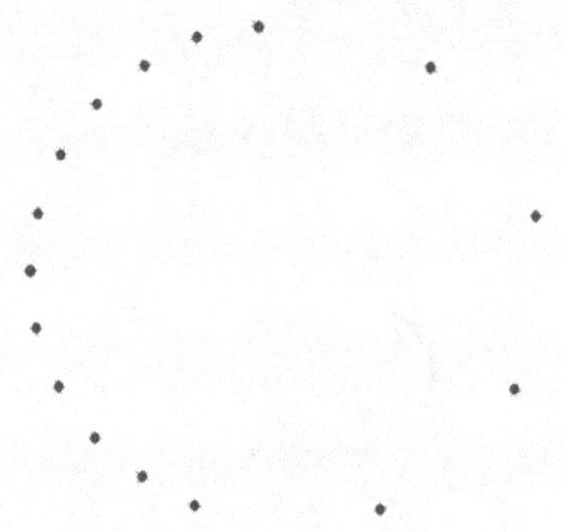

Methane is one carbon atom and four hydrogen atoms, and it has 16 main particles.

Many unrepresented tethered wave portions are helping to join one atom to another.

Assuming an additional 2 pol between atoms, a molecule of methane has very roughly 26 pol in circumference.

The formation of methane, I believe, is far from an explosive reaction. A relatively slow arrival of hydrogen atoms provides greater opportunity for bonds to form and to settle between adjacent atoms.

With an additional 2 pol between atoms, a molecule of oxygen gas has a circumference of 36 pol.

Water, 2 hydrogen and 1 oxygen, has 18 main particles and comprises of 3 atoms. Adding 2 pol between atoms gives it a circumference of 24 pol.

In the explosive formation of water, tethered wave portions, flux lines and main particles are bombarded. From an end of an atom of oxygen, some wave portions and flux lines reach out and join with either hydrogen atom. Some lines break, and others remain unsecured. Flux lines are free to grow, and they may reach other molecules.

water, H_2O

Atoms are also able to bond by being as the next diagram.

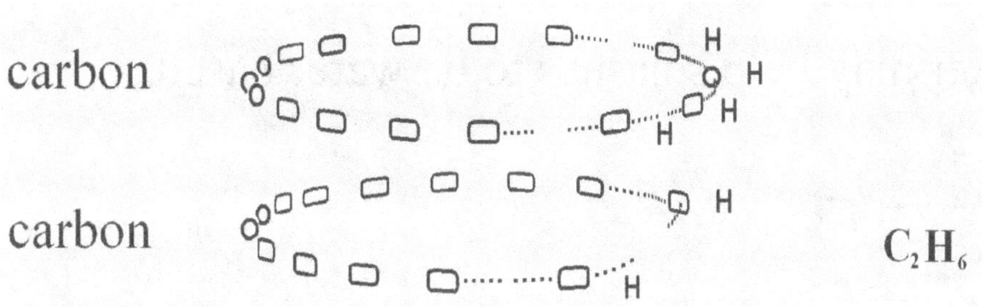
carbon
carbon
H
H
H
H
H
H
C_2H_6

The upper ring has some similarity with methane. The lower ring is under some influence of the upper ring. These two layers form ethane. It is a member of the alkane family.

More additions of a carbon atom with two hydrogen atoms can be bonded beneath this structure. These larger structures would be other alkane family members, and they are helped to exist by corresponding gaps.

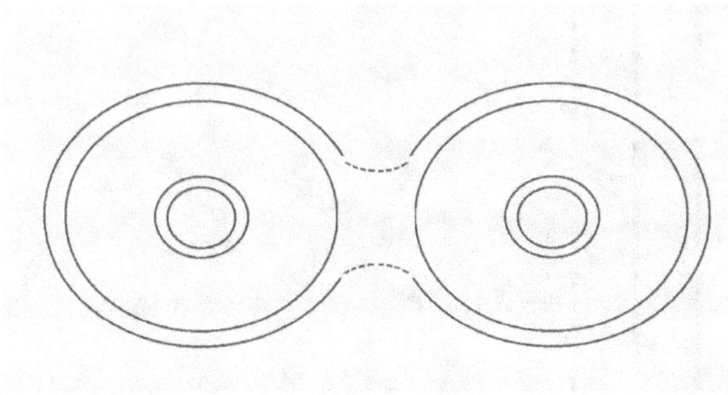

Two atoms of iodine bond by having matching gaps. Robust tethered wave portions reach a limited distance.

A group 6 gap is about twice as large as the gap of an atom of iodine. An atom of hydrogen trying to fill a group 6 gap involves flux lines of some length. These flux lines fail to compete with the thicker and denser lines of better fitting atoms.

Electricity from Chemicals

This section gives some explanation for an early battery that uses dilute sulphuric acid.

Water often has impurities. "Better safe than sorry." When adding acid to water, avoid fumes and add acid to water and not the other way round. Slowly adding the acid may be helpful. A strong acid is able to cause a violent exothermic reaction that may cause splashes. Acid to water may help splashes to be water based.

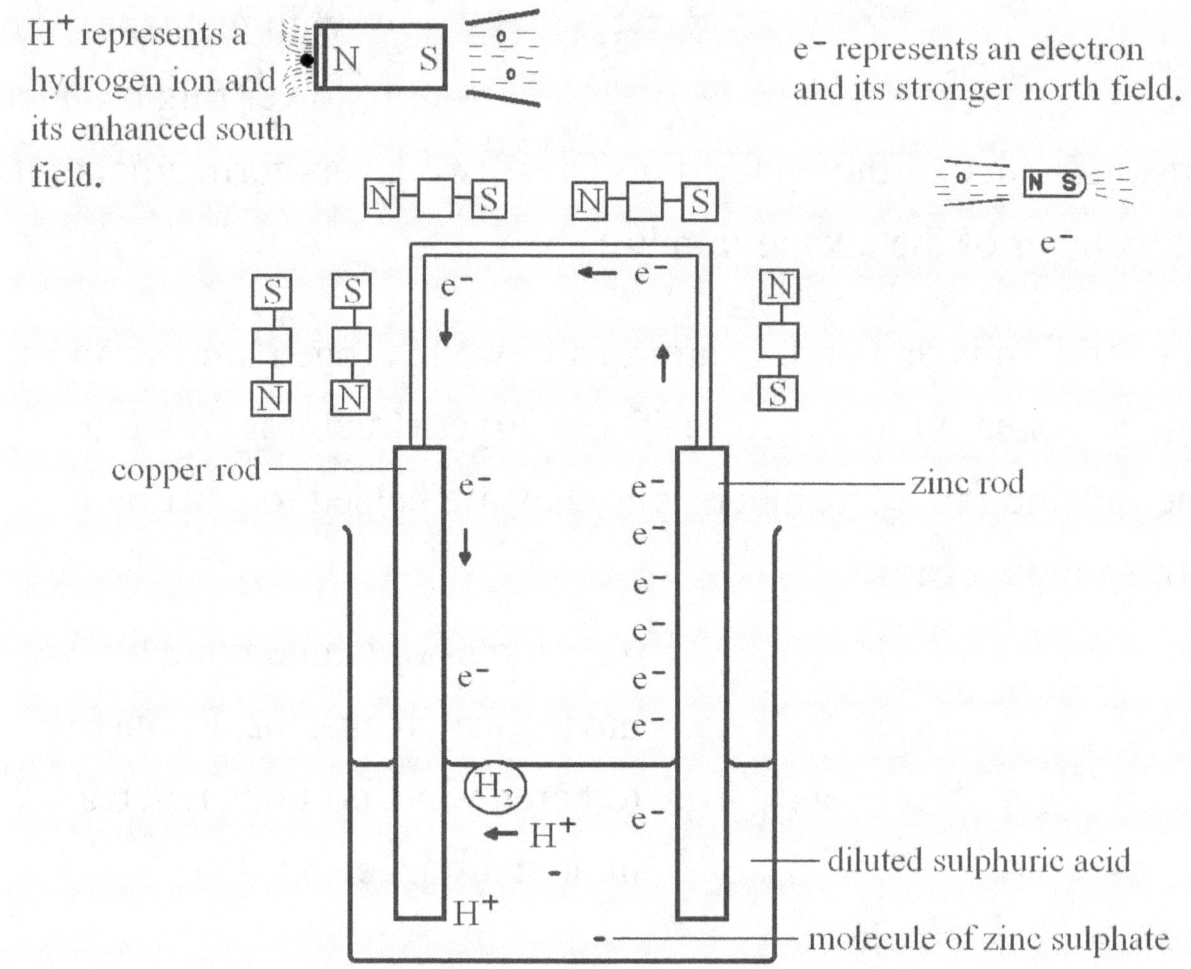

H+ represents a hydrogen ion and its enhanced south field.
N S
e− represents an electron and its stronger north field.
N S
copper rod
zinc rod
H2
H+
e−
diluted sulphuric acid
molecule of zinc sulphate

Natural $H_2SO_4 + Zn \rightarrow ZnSO_4 + 2H^+ + 2e^-$

Some atoms of zinc leave the rod to combine with negative sulphate ions. At the zinc rod, electrons are left behind.

Atoms of hydrogen are missing electrons. They are positive hydrogen ions.

An atom of copper has an outermost line of less length than an atom of zinc. A cluster of copper atoms is a tighter structure than a cluster of zinc atoms. A cluster of copper atoms has a larger number of broken tethered wave portions compared with a cluster of zinc atoms. When a tethered wave portion breaks, there is a creation of two free ends. South ends are positive and sometimes attract electrons. North ends are negative, and some north ends attract positive hydrogen ions.

Hydrogen ions are attracted to the copper rod.

Some flux lines in the wire are orientated by the positive hydrogen ions. Electrons are attracted from the zinc rod. They travel along the wire, and they arrive at the positive hydrogen ions at the copper rod.

Protons have a stronger south visible circling direction. The number of tethered wave portions and also flux lines leaving each end is not equal. After the arrival of an electron, the lines are more balanced in number.

Helped by length, two joined atoms of hydrogen has lines with some stability, and it is a gas particle.

The Great and Illustrious Wu Experiment

Associated with this experiment are Doctor Chien-Shiung Wu, a brilliant and encouraging female physicist, and the Nobel prize winners Tsung-Dao Lee and Chen-Ning Yang. This experiment has been simplified in this book.

In the passage of time, an atom of the isotope cobalt-60 loses a closely held electron. It turns into nickel-60.

Atoms of cobalt-60 are cooled to near absolute zero, near -273 ° Celsius.

At very cold temperatures, fewer expanded portions are travelling. Flux lines and small particle rings are less disrupted, which helps the flow of electricity.

In the diagram on the next page, three atoms of cobalt-60 are held in a very cold electromagnetic field. The single diagram represents three separate occasions of alignment, and it is not to scale.

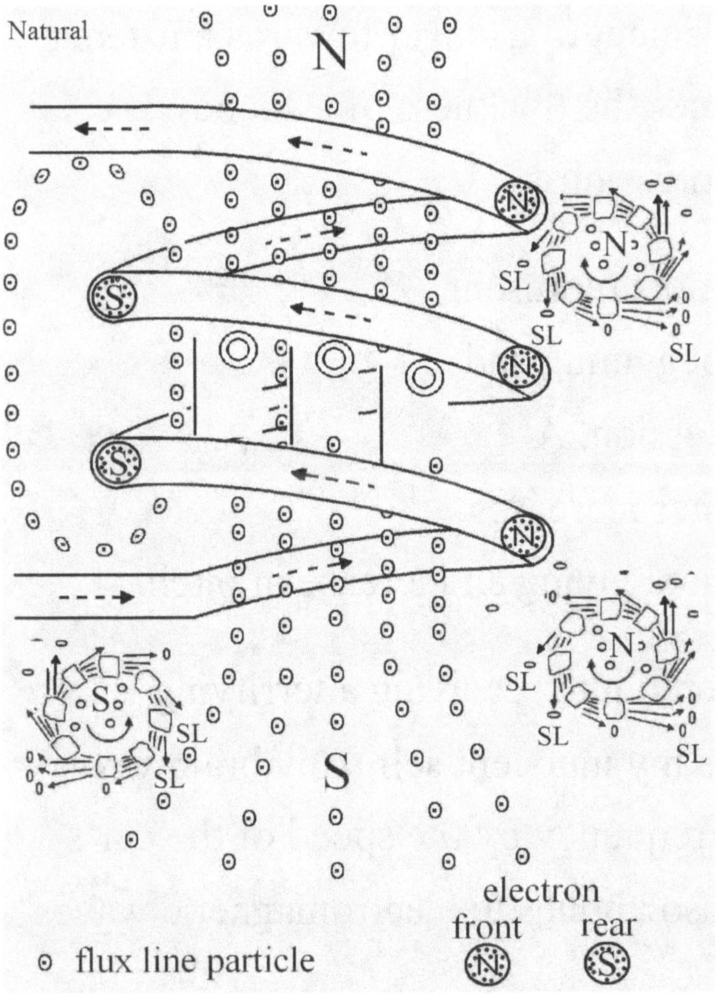
Natural
N
S
SL
electron
front
rear
flux line particle

An electron from each held atom breaks free with impetus.

The majority of freed electrons leave the south end of the electromagnetic field. It is accepted that atoms of cobalt-60 have been orientated.

The electrons appear to be from the south ends of the outermost lines.

With south being at the end of an outermost line, proton sized particles and alpha particles must have originally arrived with their own north ends leading.

The electron, being flung out from the south end of the outermost line, appears to support a system where electrons have a strong field at north visible circling direction.

The Doppler Effect

The movement of a wave source makes an actual change in frequency.

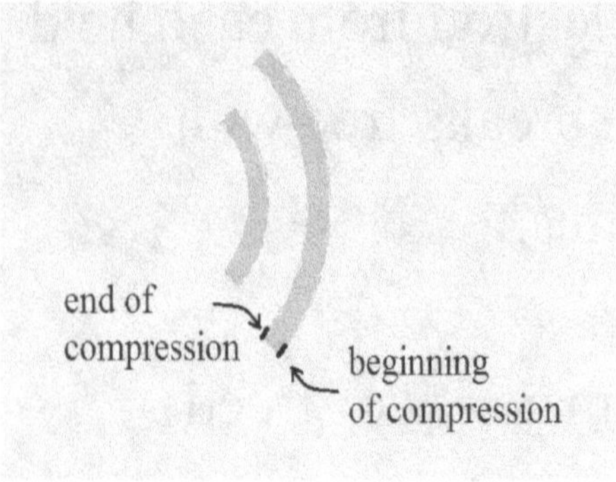
end of
compression
beginning
of compression

Movement of a singer towards a listener reduces the thickness of a soundwave compression.

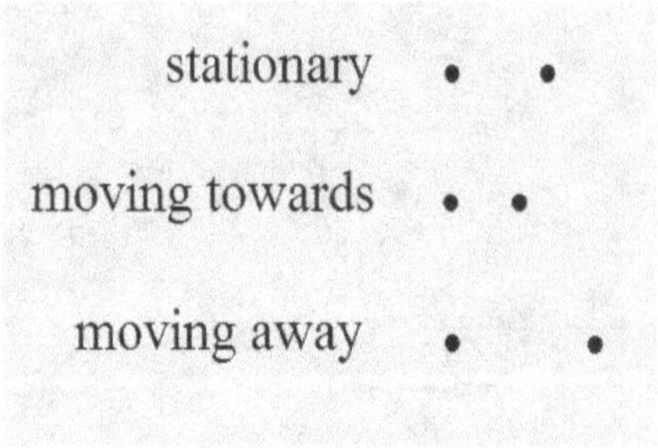
stationary
moving towards
moving away

The dots represent the beginning and ending of a soundwave compression. A source moving towards a listener increases the narrowness of a compression, giving an increase in pitch.

A cat, having had its brain altered into receiving a terrifying experience, is running towards my innocent self. Any brainwaves reaching me are increased in frequency by the speed of the cat's running. The Doppler effect also applies to electromagnetic waves.

Movement of a Listener

A soundwave is catching up with a quickly travelling listener. The soundwave takes longer to pass in this situation. Pitch is reduced.

A person moving towards an oncoming soundwave experiences the wave for a briefer period of time. The note is perceived as having an increase in frequency.

A receiver moving towards an oncoming electromagnetic wave will cause each wave cycle to pass more quickly. The electromagnetic wave will be experienced as having a higher frequency than it has. To experience a clearly orange light as being yellow light can require the receiver to be travelling at thousands of kilometres a second.

A Steady Universe

For this section, here is our initial observation: in any general direction, light is arriving from distant stars, and the light is arriving reduced in frequency by the same amount for every million years of its travelling. The Doppler effect has been used to account for the reduction in frequency.

Briefly, to fit with the Doppler effect, let's assume that the universe is expanding, and distant stars are moving away from us due to an initial explosion. We are unlikely to be at the centre of the universe.

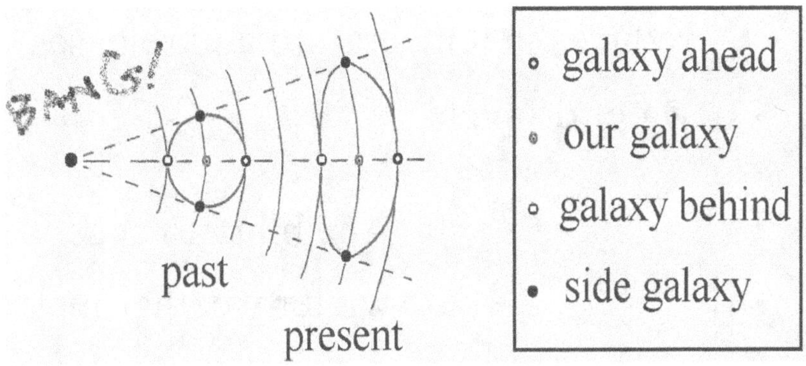
BANG!
past
present
galaxy ahead
our galaxy
galaxy behind
side galaxy

Between us and an initial bang, material only moves away from us due to differences in speed.

A long time after a theoretical initial explosion, galaxies currently at the side of our own ought to be travelling very close to parallel with us. To account for observation, expansion by movement does not fit well. Rather than involve a mysterious curving of three dimensional space, the following fits our initial observation.

Long travelling waves slowly reduce in frequency either by loss of material or, my preference, by their particles spreading to a lower concentration. Having a small effect is our own movement around the centre of our own galaxy. The universe need not be expanding.

The Blank Canvas

With effort and investment, seawater can be processed on a large scale. Deserts can be transformed.

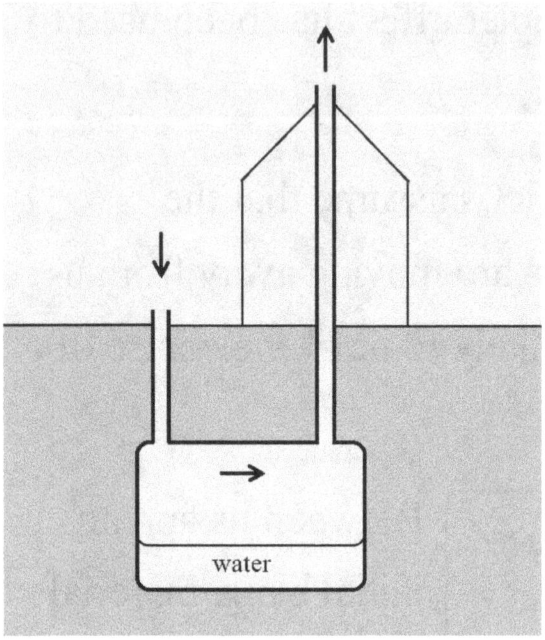
water

Hot air has greater ability to carry water than cold air.

Left is an attempt to condense water from air. Under glass, the taller pipe heats up. Upwards moving air draws air below the surface. The air reduces in temperature and releases water droplets.

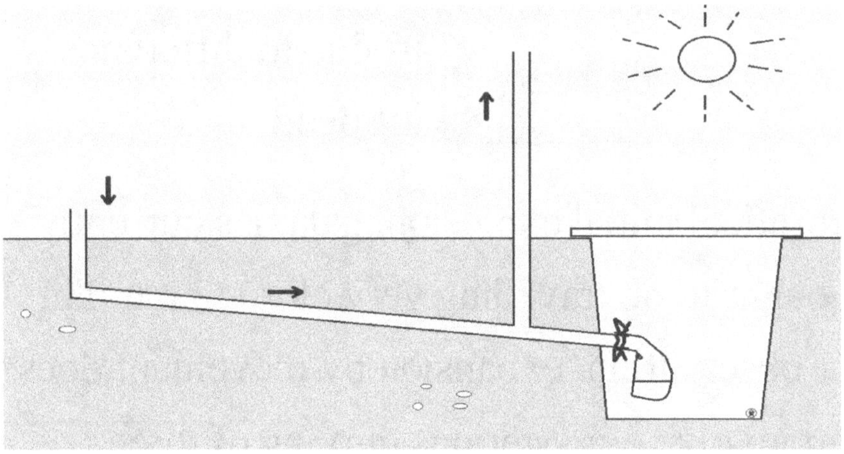

An ability to accept winter weather may reduce ground temperature.

Hopefully, water drips into the bag. The bag is inside a covered trench. Light may be reflected onto the taller pipe. The condensing chamber is going to have to disperse arriving heat. It can be composed of a four metres long gently sloping metal pipe. Burying a large metal sheet with this pipe will help conduct heat away. I offer no guarantee that it will work at all, but it may improve crops by warming the ground!

Feel free to experiment with your own version.

Frozen Water

A volume of liquid water is having its temperature reduced. At around 4° Celsius, some water molecules attract others. From these tighter areas, flux line particle structures are sent out. This material reduces the rate of the cooling.

At around zero degrees Celsius, the molecules are held in a fixed position. They take up more room. The molecules are ‘holding hands’. Alongside each other, they join and form large shapes.

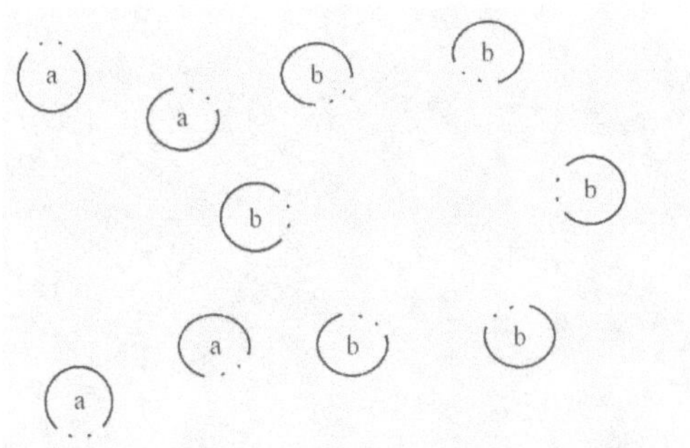
a
b
b
a
b
b
a
b
b
a

Flux lines can be imagined holding the molecules in place. Molecules marked with an ‘a’ are at approximately the same level as those marked with a ‘b’.

Repeating patterns are likely with polarity and flux lines.

Ozone

A molecule of ozone is three atoms of oxygen. Under a small amount of bombardment, molecules of ozone collapse and form molecules of two joined atoms of oxygen. Often, the ozone to oxygen change briefly leaves an oxygen atom with unoccupied flux lines. Occurrences of this event are able to block a wide range of electromagnetic waves.

Ions Travelling in a Magnetic Field

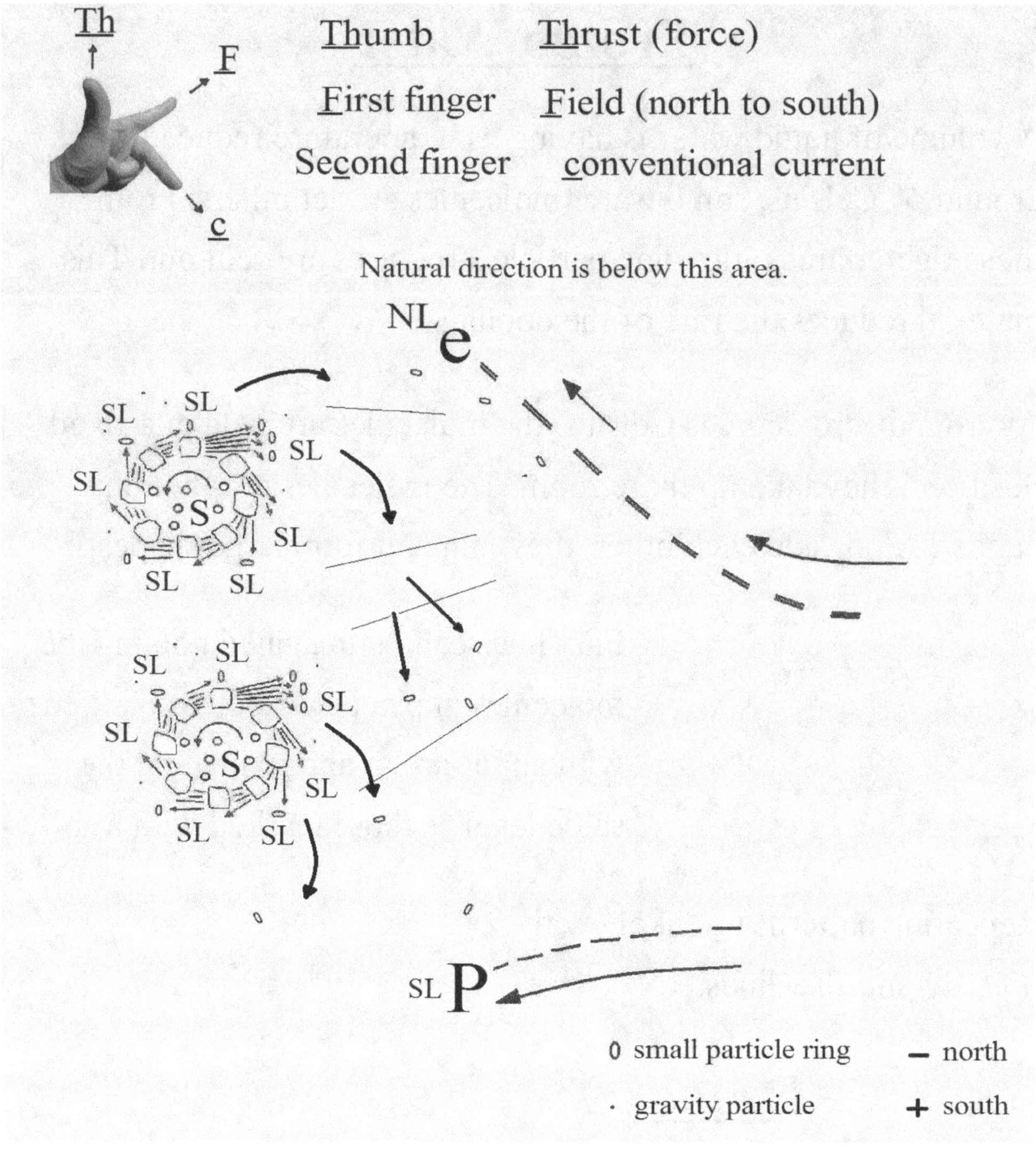

Th
F
c
Thumb
Thrust (force)
First finger
Field (north to south)
Second finger
conventional current
Natural direction is below this area.
NL
e
SL
S
P
0 small particle ring
· gravity particle
– north
+ south

The above includes an end view of two flux line particles. They are a small part of a collection of parallel flux lines.

Fleming's left-hand rule is for electrons moving in a magnetic field. To convert a Fleming's hand rule to natural current, have current flow travelling towards the palm.

Small particle rings are leaving from the sides of flux line particles. Their direction of release at the outside edge of the magnetic field dominates.

Electrons with a stronger north end experience some attraction. In the arrangement in the diagram, electrons are deflected upwards.

For a measurable effect, electrons must arrive at a perpendicular angle to the flux lines; in this way, electrons are oncoming against many leaving small particle rings, and timing is less involved in the occurrence of collisions. To a very limited extent, it is the difference of walking down a motorway against oncoming traffic to the alternative, walking across at an angle.

For protons heading into a magnetic field at a perpendicular angle, their south ends are repelled. To the right of the field in the diagram, some small particle rings are travelling downwards, and the proton is being repelled downwards.

Moving a Wire into a Magnetic Field

To generate electricity, wires must move perpendicularly into the field. Small particle rings leave flux lines in a perpendicular dimension. Small particle rings are most likely responsible for meeting an oncoming wire and releasing electrons.

With Fleming's right-hand rule, the thumb points in the direction of movement of the wire. 'C' is conventional current direction. (Conventional current is recognised to be an early mistake.) Natural current travels in the opposite direction.

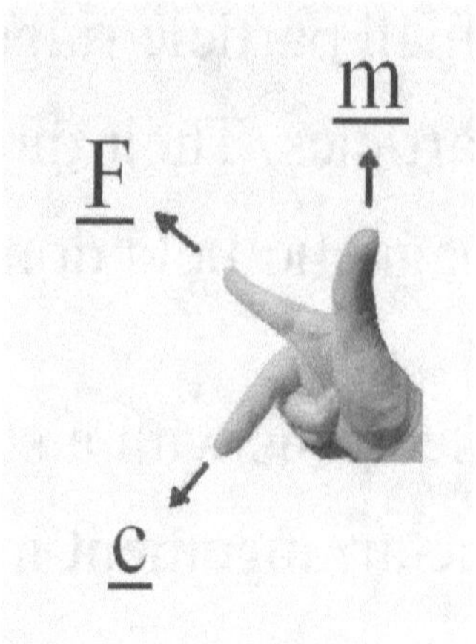
m
F
C

'F', field direction, is north to south of two involved magnets or flux line particles.

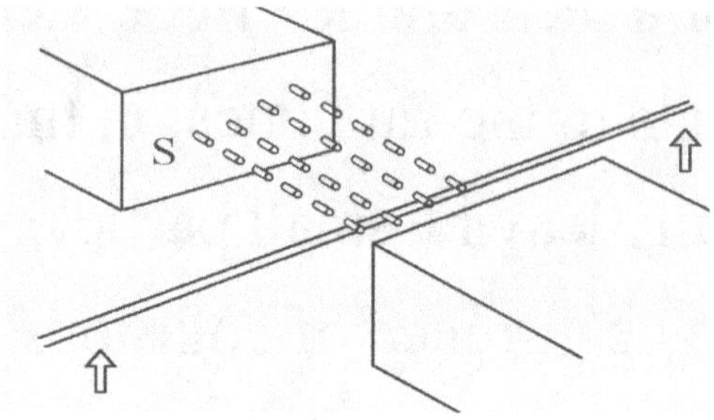
S

Small particle rings from the outside edge provide the main direction of travel.

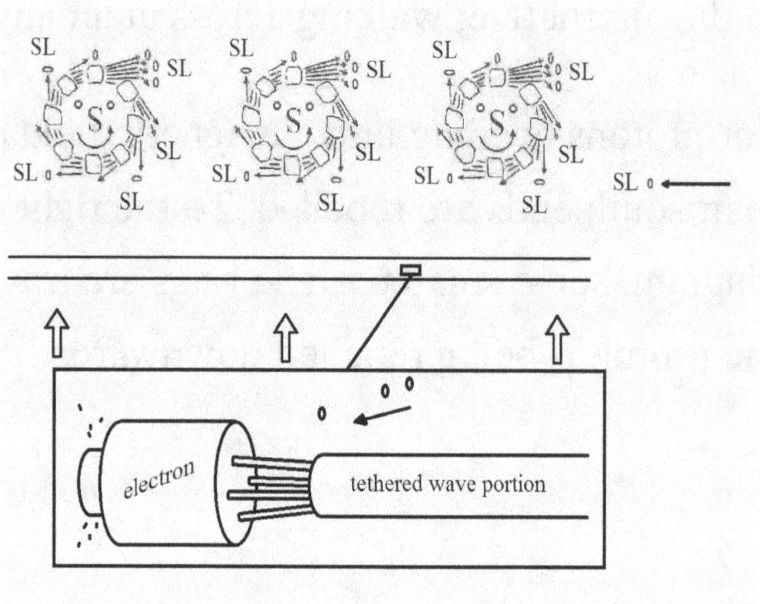
SL
S
electron
tethered wave portion

In the diagram: the wire meets small particle rings travelling to the left; an electron is going to be struck on its north side, and it is about to be released; in this formation, electrons are released from the left side of structures.

Flux lines replace the electrons. Alignment works its way along the circuit.

Aided by disruption, extracted electrons sometimes pass their docking areas. These electrons follow a route of increasing alignment. They travel around the circuit, and they return to the docking areas with missing electrons.

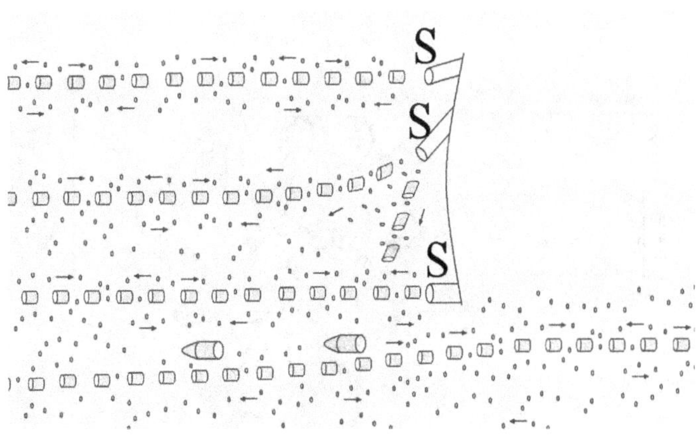
S
S
S

Quickly moving magnets, orientated as on the previous page, send flux lines over and inside a coil. The coil consists of insulated copper wire. Four docking areas are represented.

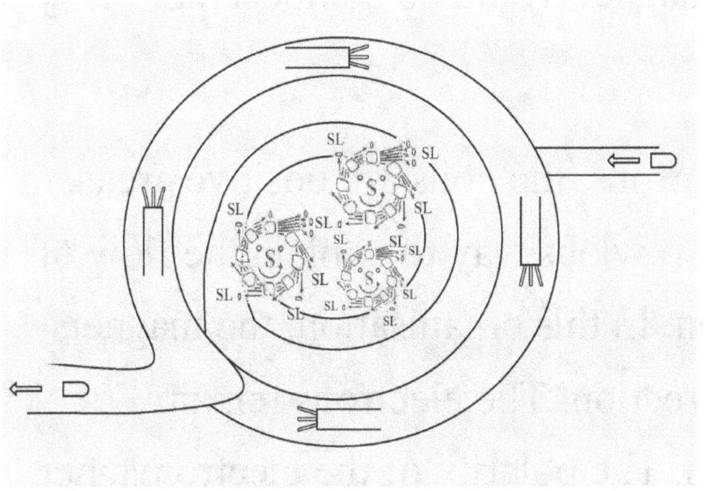

Electrons move in the direction of the arrows.

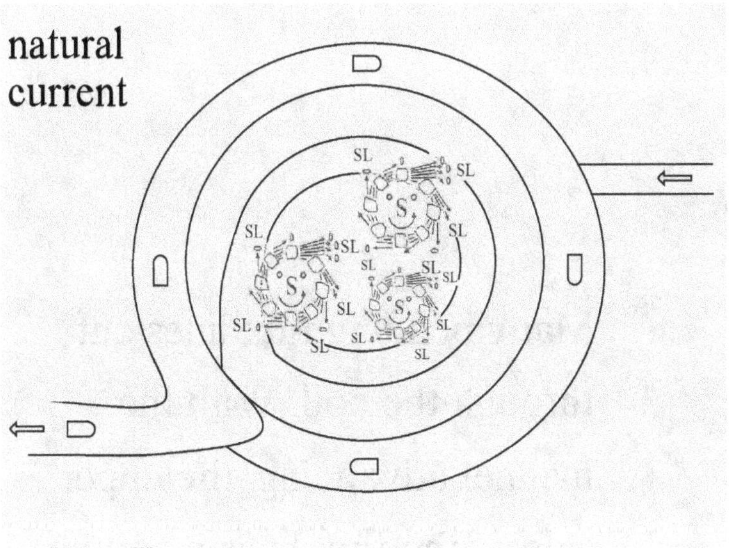
natural
current

In the next diagram, the coils are wound in the same direction. Pairs of magnets cover the coils briefly. Electrons are extracted from the one side of structures in the wire.

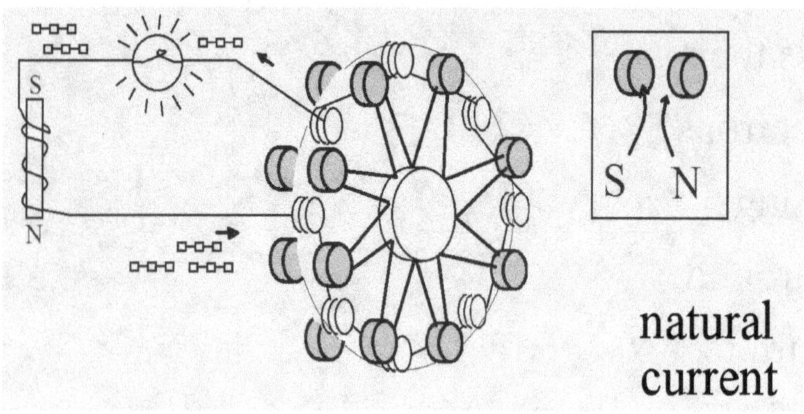
S
N
S N
natural
current

Electrons travel around the circuit. They are attracted by the increasing alignment that is represented by the groups of three boxes.

The coils and also the electromagnet would be composed of many turns.

With the coils losing some electrons, the coils are positive areas. Without alignment, there is no obvious way to explain the flow of electrons being in one direction. In this organisation, the magnets are able to revolve in either direction. The electrons remain travelling in the same direction. The polarity of the electromagnet, also, does not change.

<u>Moving a Magnet into a Coil</u>

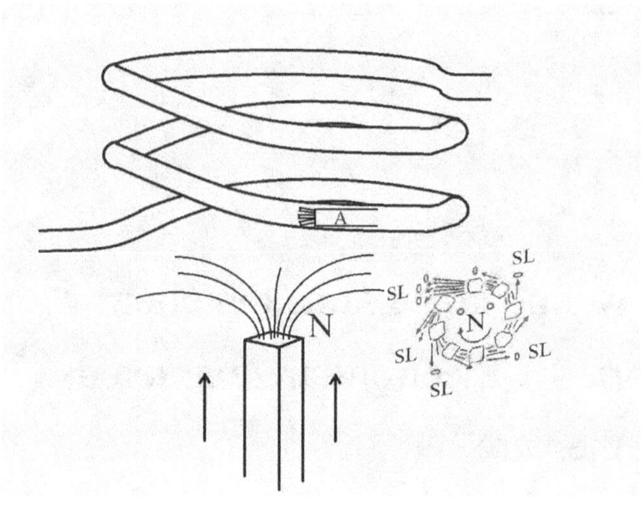
A
N
SL
SL
N
SL
SL
SL

Many bending flux lines cut through the coil. With the magnet advancing, the upper side of the flux lines meet the wire and extract electrons. Broken tethered wave portion ‘A’ is missing an electron.

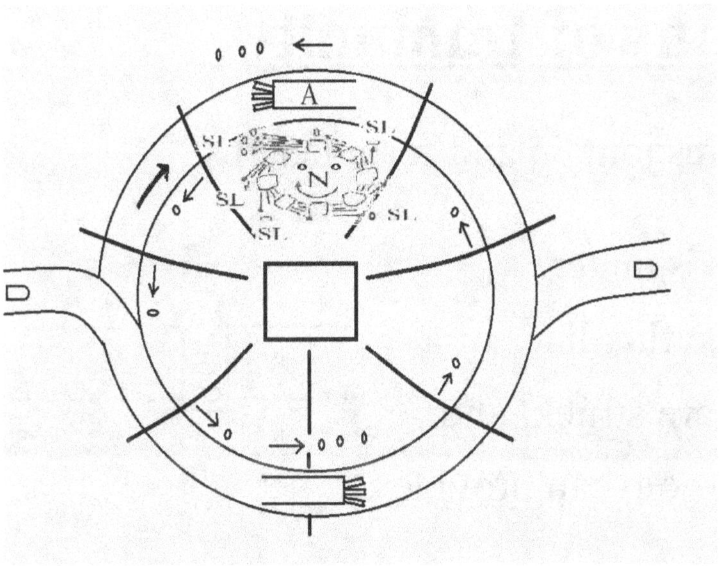
A
N
A

The north pole of the magnet advances quickly. Small particle rings travel into the wire in a counterclockwise direction. Electrons are sent in a clockwise direction (natural current).

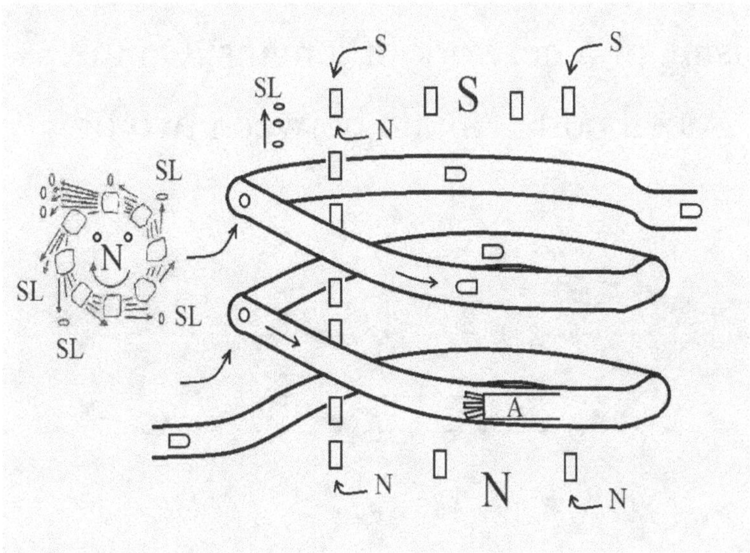
S
S
SL
N
S
SL
N
SL
SL
SL
SL
A
N
N
N

Electrons forming the current travel along the coil, and they send south leading small particle rings upwards inside the coil. The small particle rings attract flux line particles.

A north field is created at the bottom of the coil.

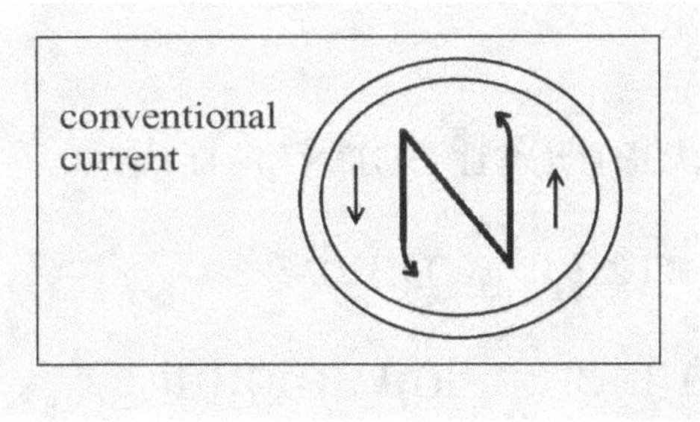
conventional
current
N

Conventional current can be misleading. An advancing north pole sends electrons clockwise, not counterclockwise as suggested by this diagram.

Withdrawing the north pole of a magnet quickly also generates a current. The lower side of bending flux lines reach the turns of wire first. Along the coil, electrons come to travel counterclockwise. A south field is created on the magnet side of the coil.

A Philosophy of Immunity

In the human body many atoms join up and form lengths.

The join between two lengths is under continual bombardment. Some flux lines of a join break. Broken flux lines are struck, and they are directed outwards. A join can provide a level of attraction.

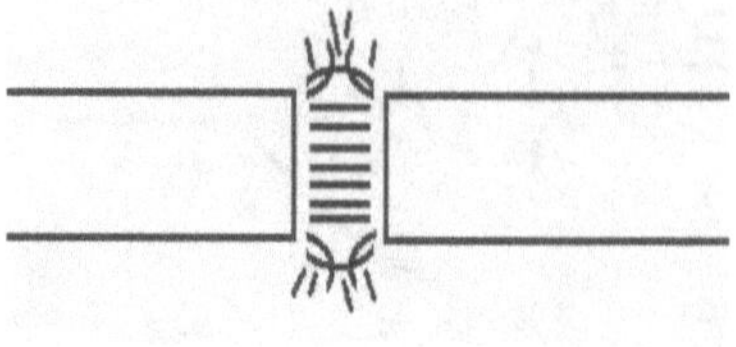

Cells have a surface comprising of a network of lengths. On the surface there are cell ports. A cell port is a join between two or more lengths.

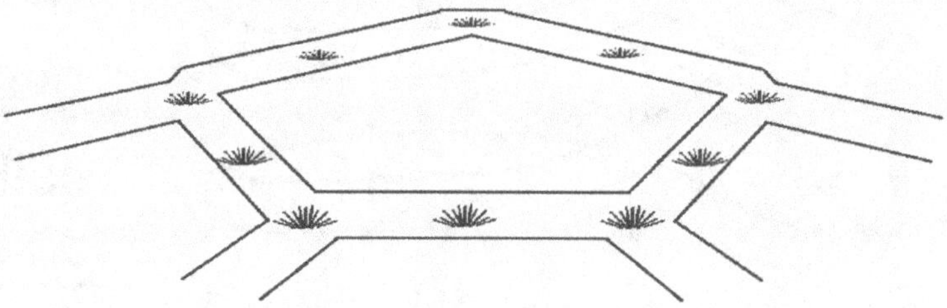

Cells of the same type include a matching configuration of cell ports. A matching configuration enables cells to join together.

A virus structure locks on to a group of human cell ports. An invading spike may save a virus from requiring a strongly matching formation. When a virus gains an increasing amount of connectivity with cell ports, joins along the surface of a human cell become weaker and fall apart. Lengths of material from the human cell are attracted to the virus cell. With enough material the virus cell replicates itself.

When a human cell collapses, a portion of it escapes being used by a virus. Escaped material contributes to the building of defensive structures.

An effective vaccination provides a defensive structure or something resembling it. A defensive structure attaches to a virus structure and disables the invader.

The defensive structure, being attached, has added stability. The lack of movement helps to draw in lengths. Drawn in lengths attach and begin to replicate the defensive structure.

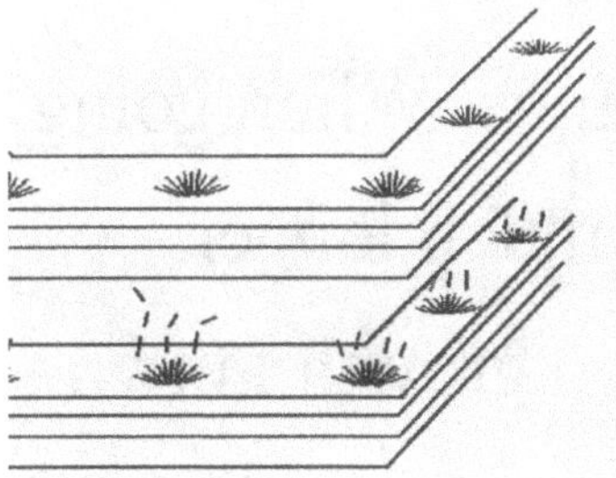

Newly arrived lengths bond more strongly to each other than the defensive structure being copied. The field of the newly arrived lengths becomes strong enough to cause separation from the structure being copied. With a suitable supply of material, the number of useful defensive structures rapidly increases.

During the initial phase of illness, lengths tend to join onto larger structures in low numbers at a time. Vaccines, designed to be accepted during health, tend to be well suited to the initial phase.

During prolonged attack by a virus, released lengths become common. Before attaching to a vaccine structure, lengths attach to each other more frequently. Different structures can result. In this phase, cells can become clogged with our own material. A clogged cell often fails to join with newly formed cells. Sometimes, an invading cell is smothered.

One Becomes Two

Washing-up liquid undermines the attachment of grime, and it also joins ends reducing the ability of grime to return to sticking.

A solution including washing-up liquid and cells creates a long jelly-like double helix. Rather than amplifying an existing structure, the washing-up liquid is breaking up cells and joining lengths. Two main columns attach to each other at their sides, and they grow. While the structure grows in length, the area farther out from the centre is more accessible. This farther out area gains material and length at a faster rate. The quicker growth is also attached towards the joined edges of the two columns. When going up a winding staircase, we are able to recognise a much longer outer edge than the inner edge. This kind of uneven growth in two connected columns results in a double helix.

As an internal structure of a cell, the double helix would require a mechanism to interpret it. In its mapping, I believe scientists have found a way to break apart and record attributes of the surfaces of cells. Although many of us have seen the double helix in artistic interpretations, it has never been photographed inside a human cell. The structure is theoretical and open to contest.

Cells include the simple structures which largely define our other cells. Most of the surface of a cell is composed of these simple defining structures. Being on the surface of a cell makes duplication easier. Being on the surface also helps a defining structure to have an influence.

On the surface of cells, there are also connecting structures. Each type of cell has its own connecting structure or connecting structures. The field of a cell is a result of the simple defining structures and also the connecting structures.

When a cell dies, its connecting structures tend to survive and are cast off. Freed connecting structures are knocked about until they join another cell of the same type. Unless we are experiencing either a stage of growth or interfering factors, connecting structures ensure that cell multiplication is governed by cell death.

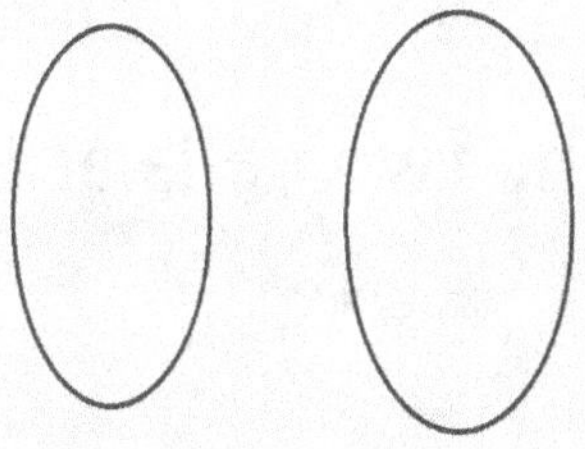

A cell can be seen to enlarge.

Lengths arrive. From dead cells, matching connecting structures arrive. Together, lengths and matching connecting structures duplicate the surface.

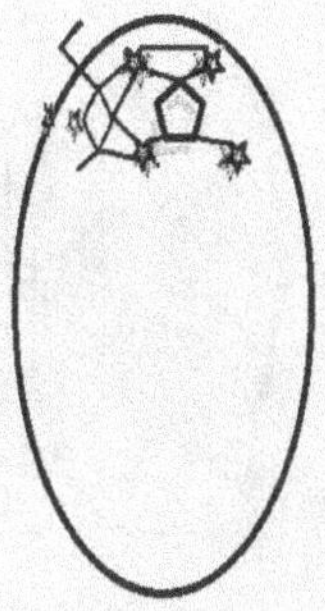

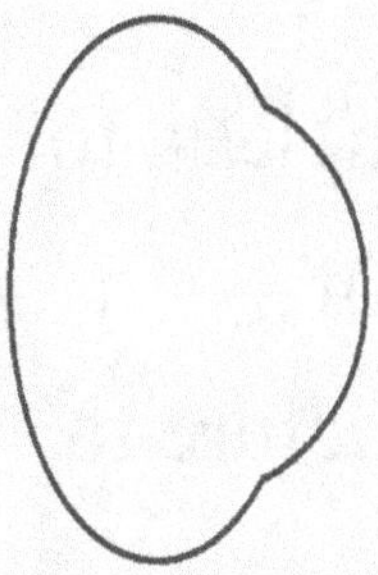

With additional material, field intensity increases for a portion of the secondary surface. It pushes farther out. Flux lines manage to maintain a grip.

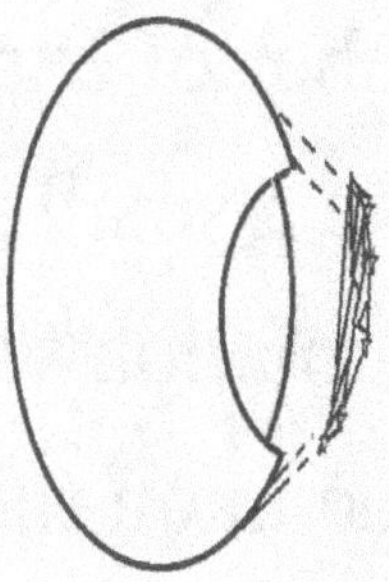

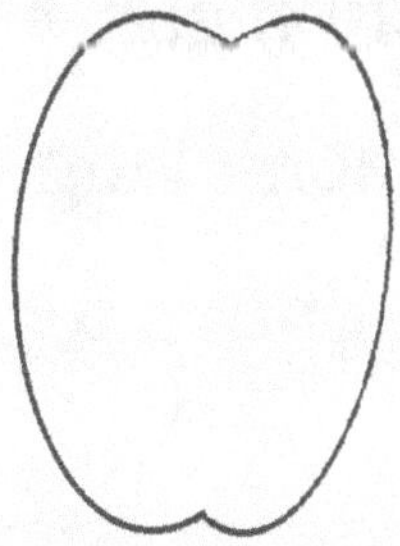

The secondary structure is held open by the primary structure.

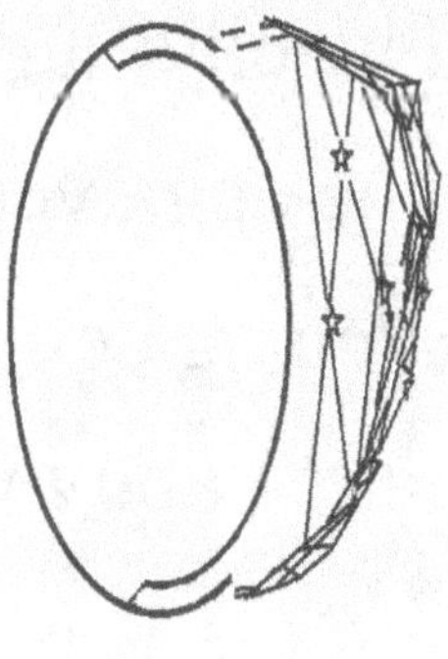

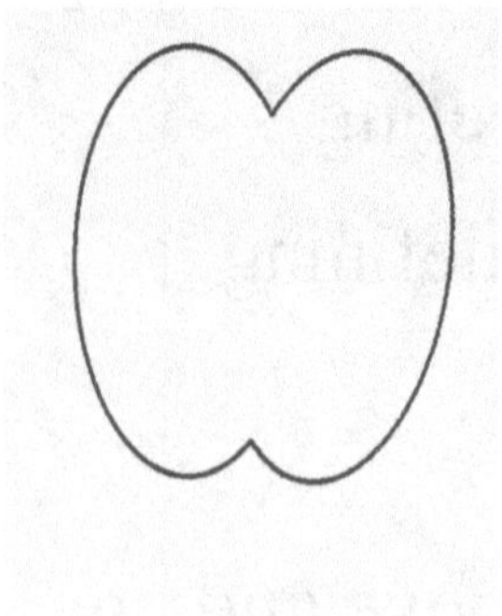

The secondary structure increases its field. It springs farther out. Pressure on the primary structure reduces. The secondary structure begins to close up.

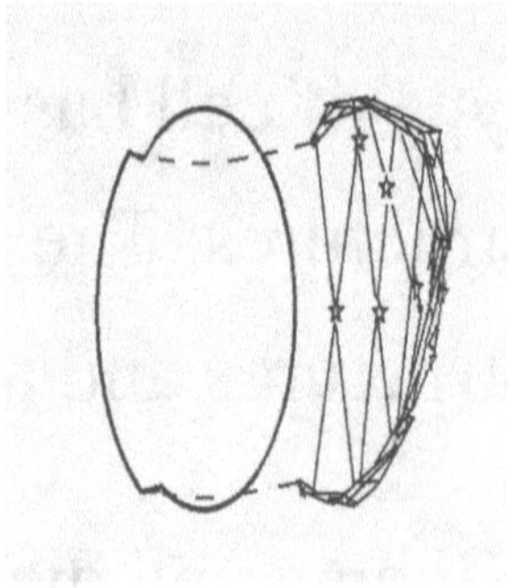

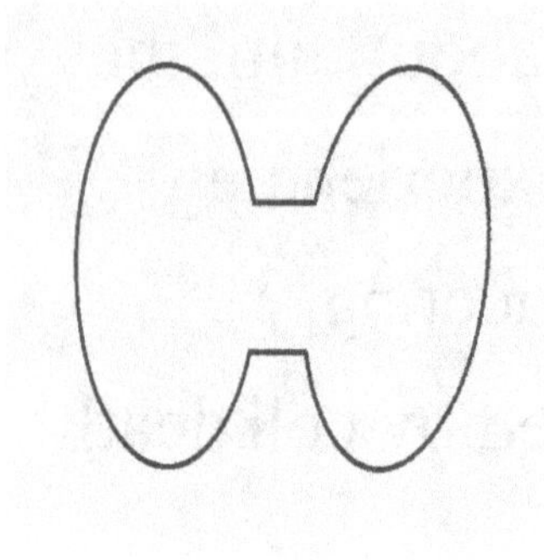

With the aid of flux lines, material continues to transfer.

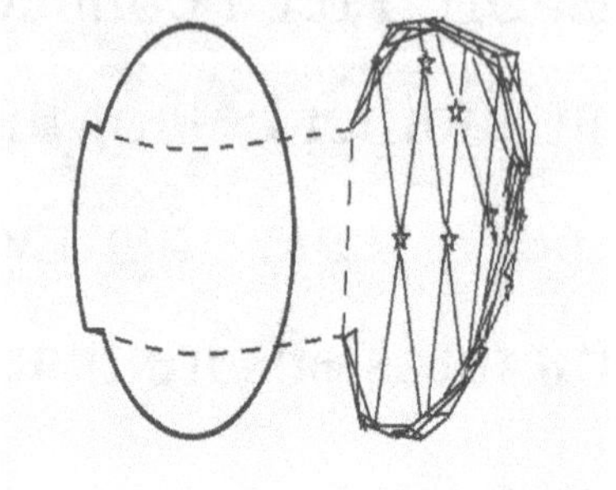

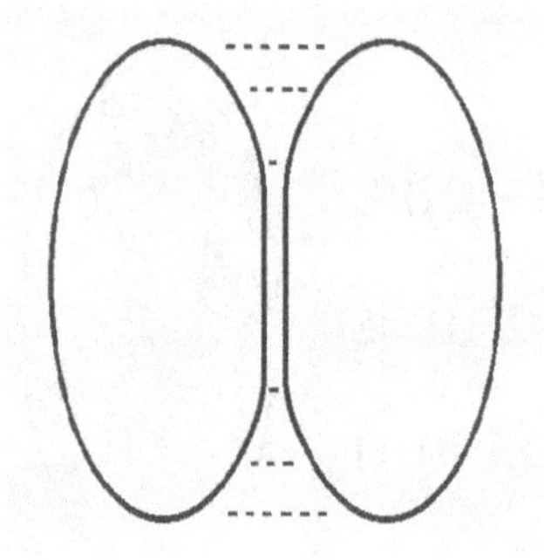

A level of separation occurs when the new cell is complete. By way of flux lines, the two cells are cosily 'holding hands'.

Hair roots follow the system using a fixed number of connecting structures. Hair is less restricted.

The idea of a double helix being inside a cell has led to a huge number of experiments and important advances. Nonetheless, to make a small substitution inside a relatively long string of information would, I believe, require the introduced structure to home in on the portion to be replaced.

At a cell's surface, the homing in and the substituting action can all be achieved in the formation of a secondary cell. By achieving matching cell ports, an introduced structure is able to be drawn to an area above a primary cell.

One Good Turn

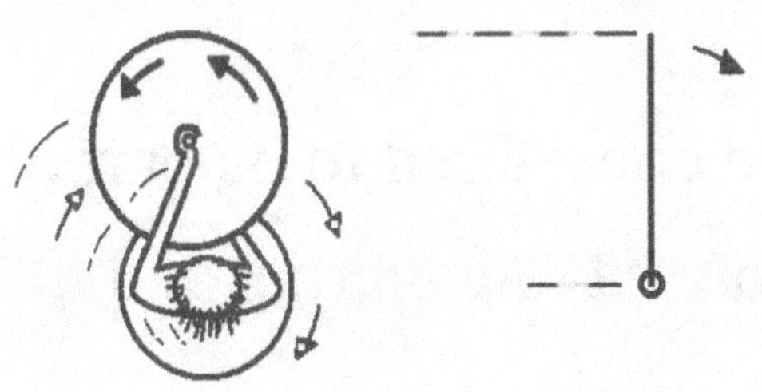

My imaginary self, a fearless daredevil and shockingly handsome, is standing on a turntable, and I am holding a quickly turning disc. The outermost edge of the disc is turning counterclockwise. Every force is said to have an equal and opposite. One force is occupied with turning the disc; this force is continually changing direction. An opposing force is also changing direction.

To my own centre of turning, the outermost edge of the quickly turning disc is more distant than the near edge. On the longer radius, the opposing force has the greatest leverage, and it causes me to turn in a clockwise direction.

Rotation of Earth

My imaginary self has an imaginary partner. We enjoy digging holes in the garden. One never knows what one might find. I have found several rusty nails. My father, Popsy - I like to call him, proudly keeps them in a jar for future use. My imaginary partner and I have dug a well in the garden. When the well was first dug, my partner cheerily dropped a coin down its centre. In the rain, I searched for the coin. Rest assured, we still talk to each other.

My partner in being a bit of a spendthrift highlights an important issue. In order to fall over four metres directly into the water as it did, the coin must have kept up with the turning of Earth.

The diameter of the equator is taken to be 12,756 km. At the equator, in taking one day to make a revolution, the surface is turning at approximately 464 metres per second.

The coin, which fell down the well, can be considered to be part of Earth's surface. It travelled with Earth from having inertia, and it also travelled downwards due to gravity.

When an object moves with an increase in speed, groups of spherins inside spheres move to the rear. For an object, the position of spherins is connected to inertia.

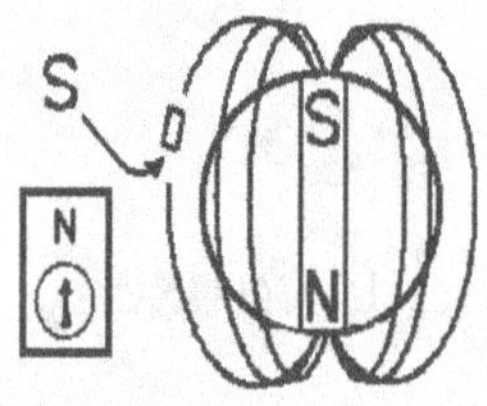
S
N
S
N

The physical magnetic south is in the Northern hemisphere. Circling particles send out expanded portions and attract Earth into turning.

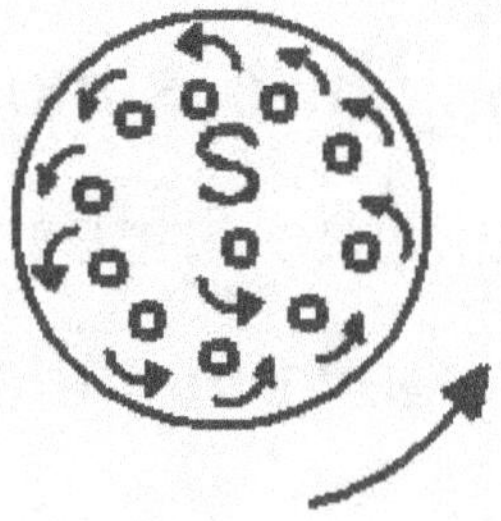
S

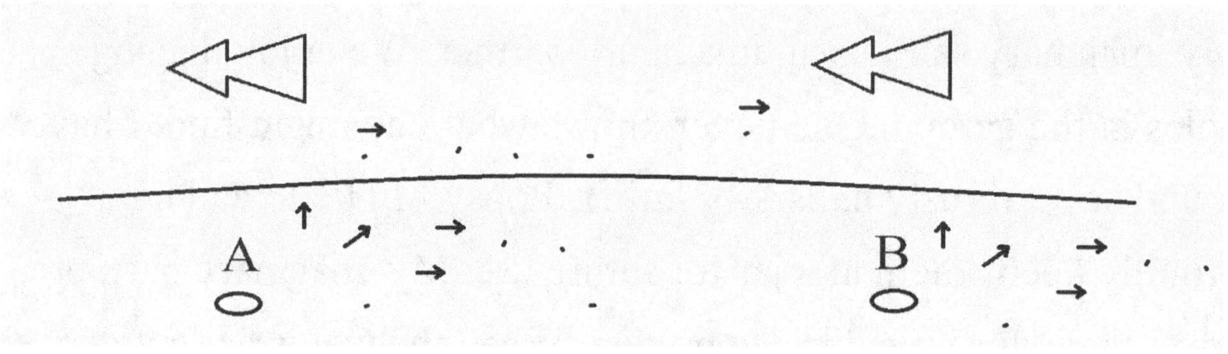
A
B

The spheres of pebbles 'A' and 'B' contain more spherins on the right-hand side than the left. Along the equator and parallel with it, material in the surface of Earth is sending out expanded portions, more in one direction than the other. A sphere shape, being a collection of circles, helps to maintain the bombardment.

Long ago, expanded portions caused Earth to begin turning. They continued applying a force. They increased the speed of rotation until resistance became too hampering.

The Circling Directions

People investigating the system may find this section of some interest. This section is not essential reading.

While developing the system, there has been some contention over polarity. I believe there to be two main deciding factors: the Wu experiment giving electrons a strong north polarity; and also the attractive reaction causing Earth to rotate, leading to the idea of spheres at south visible circling direction being with counterclockwise rotation.

The alternative has north visible circling direction as counterclockwise. We can look at how it fits.

In the 'Electromagnetism' section, to create the same polarity, north leading rings must travel up through the centre of the pictured coil.

In 'Ions Travelling in a Magnetic Field' the pictured diagram would have north leading rings travelling upwards. Protons are now attracted downwards, and electrons are repulsed upwards.

In 'Moving a Wire into a Magnetic Field', north leading small particle rings come from the left, and they strike the outer south side of electrons to release them.

This alternative arrangement requires a convincing reason for the physical south end of Earth to rotate counterclockwise.

Redirecting Gravitational Force

Some years ago, a trapdoor in my mind was falling open. A weight on the door immediately fell through. "Gravity is indeed instantaneous," I declared to myself. The following experiments prove otherwise.

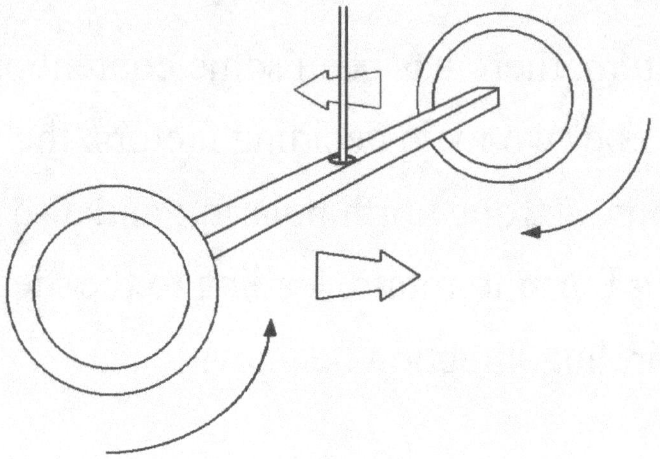

Central is a vertical rod. A horizontal beam is free to rotate in the horizontal plane. A wheel is attached to each end of the beam.

The wheels are free of the ground. The nearest wheel is sent turning quickly in a counterclockwise direction.

When stationary, gravitational force would be pulling the nearest wheel downwards.

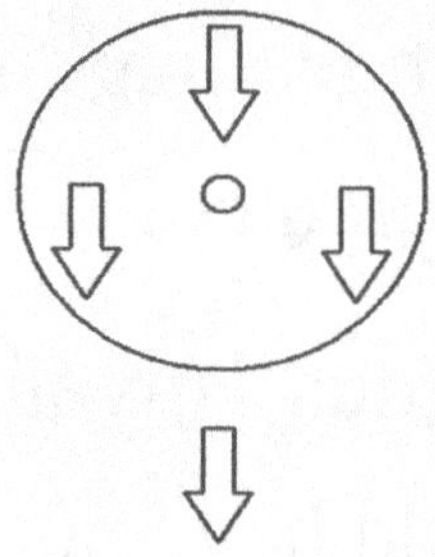

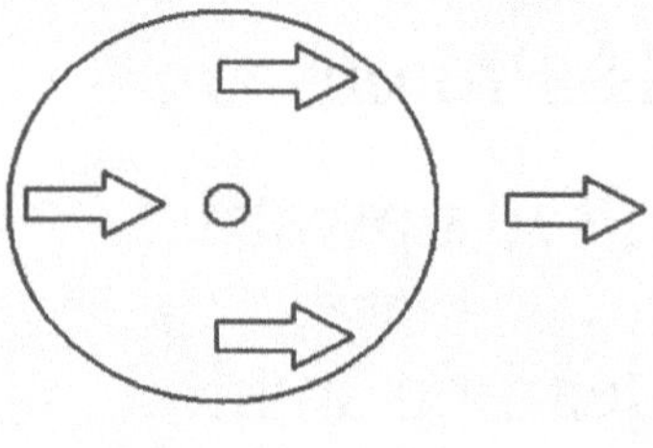

The wheel is turning quickly in a counterclockwise direction. After a quarter turn, gravitational force is pulling the wheel to the right.

To look down on the experiment, the horizontal beam reacts to the force and turns counterclockwise. (Care must be taken in the direction the second wheel is sent turning.)

The experiment with the wheels is an adaption of an experiment that I saw on the internet. The following resembles the internet experiment.

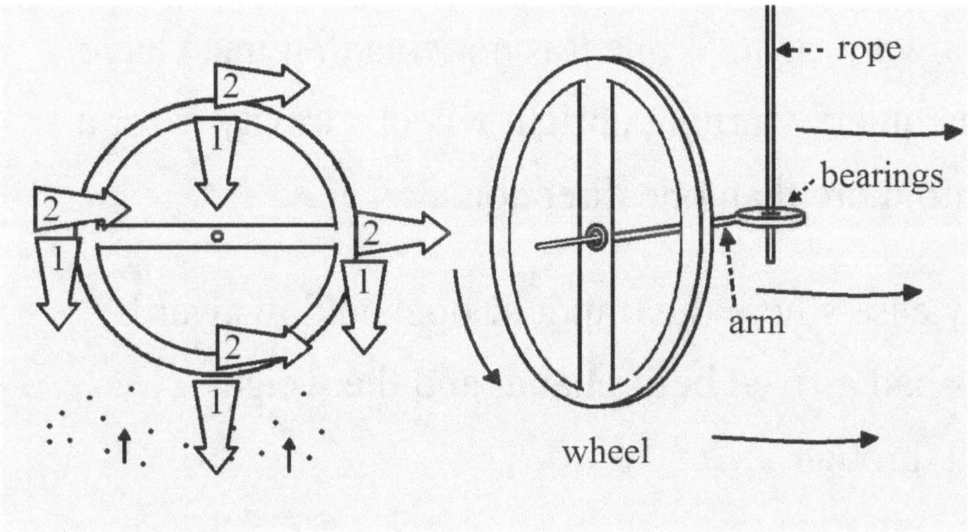
rope
bearings
arm
wheel
2
1
2
1
2
1
2
1

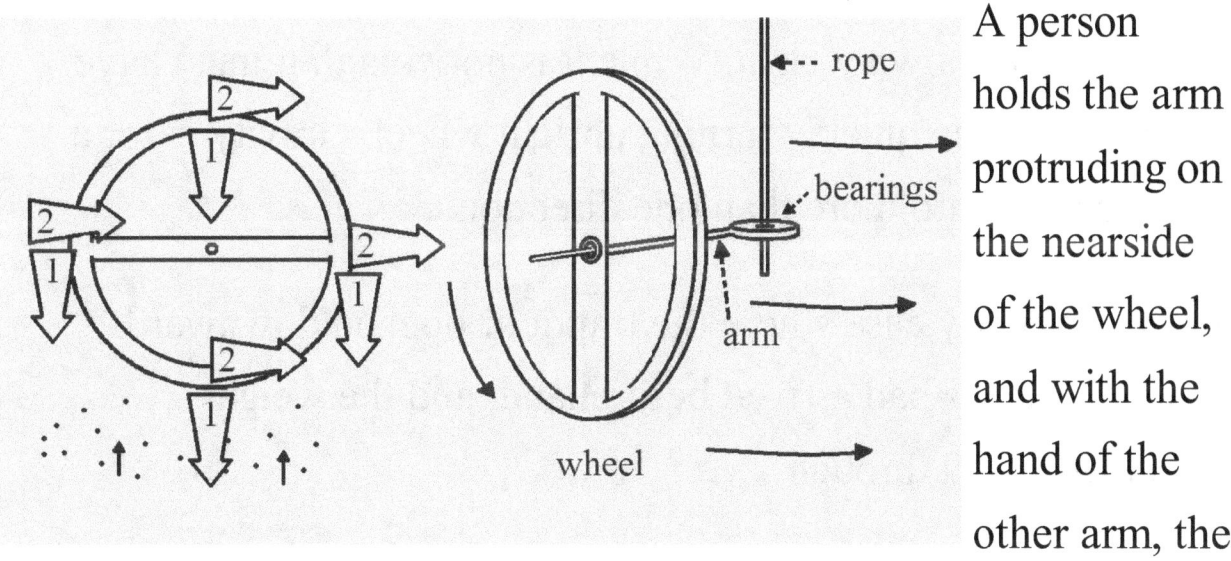

A person holds the arm protruding on the nearside of the wheel, and with the hand of the other arm, the experimenter quickly and repeatedly brushes the wheel. The brushing action involves a hand making a chopping motion through the air.

The arm, which the wheel rotates around, is light in comparison to the wheel. A wheel being heavy near its outer edge maintains its speed for longer, and it builds up speed more readily by the brush of a hand. With the wheel rotating quickly, the protruding section of the arm is carefully released.

If the wheel rises, the arm and rope cause the wheel to tilt. Gravitational force is no longer directed upwards. Gravity pulls awkwardly on the wheel.

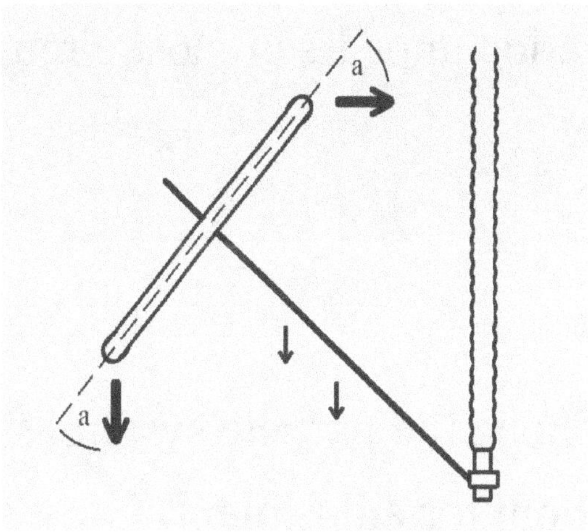
a
a

In the experiment, the wheel makes a little jump in the air and returns to being vertical. The arm spends time in a horizontal position.

The arm has its own weight. While it is horizontal, it must have support from the quickly turning upright wheel. Gravity is for a time smeared into more than one direction.

In my imaginary case where the trapdoor opened downwards, gravity particles had arrived beforehand, and the weight immediately fell through.

The system of gravity expressed by our understandably beloved Albert Einstein is, I believe, regarded as instantaneous, and it seems to involve a mysterious collapsing or curving of space between two masses. The source of its accuracy may simply be its use of a great deal of study and careful observation.

The simpler and in someways more complete system proposed by myself involves the arrival of very small particles. The particles cause a knock on effect. The effect spreads through a mass, and it takes time to complete. With the action occurring all the time, the action appears in most circumstances instantaneous.

I hope the particle system is used wholesomely. There is some room for adjustment.

www.ingramcontent.com/pod-product-compliance
Lightning Source LLC
Chambersburg PA
CBHW081051170526
45158CB00007B/1943
9781291897333